THE CLASSROOM

Preface/Acknowledgements

To me, beekeeping has got to be one of the most unique experiences available to man. Under what other circumstances can humans and insects be mutually beneficial? Not too many. Man and honeybees have been linked together for thousands of years. At first man adapted his practical everyday observations to the management of honeybees for increase. Later science joined in the study of the fascinatingly complex life of honeybees. And the fascination with honeybees, what they do and how they do it has not ceased. Even though the honeybee has had more written (except for religion) on every aspect of its life, management of honeybees still continues to be equal parts of art and science. That's probably why honeybees are so interesting for generation after generation. As soon as the beekeeper thinks he has an understanding of why honeybees do what they do under a particular situation and that he can now predict their actions, they change their response completely. Beekeeping is not only an enjoyable pursuit, but also a humbling one. Curiosity of why honeybees do what they do is stimulating and perpetuated. This curiosity begets questions and sometimes, correct answers.

Sharing those questions and answers with other beekeepers is part of what makes beekeeping fun and that's how the "Classroom" section of the American Bee Journal began. Beekeepers writing in to ask questions and share answers and experiences.

The questions and answers found in the book have been compiled from the last five years of the "Classroom," some from the 1978 version of the book "Beekeeping Questions and Answers," and some from my back files.

Ultimate thanks goes to God for allowing us to enjoy life with one of his creations, Honeybees. Next, thanks to all of the beekeepers who read the "American Bee Journal" and the "Classroom". Their continuous fascination and appreciation of honeybees are what make the "Classroom" possible. Thank you to Joe Graham, Editor of the ABJ. He has been extremely supportive over the years in allowing me quite a lot of latitude in what types of questions and answers appear in print. Without his support and encouragement this whole project would not have been possible. Special thanks to the following helpers, proof readers and friends that made this book possible: Shirley Moeller, Bertha Disselhorst, Ivan Pilkington, Terry Avise, Dave Coovert, Steve Martin and anyone else I may have forgotten.

It's been fun. Hope you enjoy this. Let's do it again sometime!

G.W. Hayes, Jr.
"Jerry"

TABLE OF CONTENTS

The Classroom

CHAPTER 1

QUEENS

QUEEN INTRODUCTION

Question - In two separate beekeeping books I've seen an explanation of placement of a queen cage two different ways.

One book says "The queen cage is suspended between the frames with the cork removed from the candy end of the cage and *with the candy end of the cage up.*" Another book says, "Be sure that the end containing the candy is *toward the bottom of the hive.*" Which is correct?

Abdul Bonaparte
Clydesdale, NE

Answer - The candy end of the queen cage should always be facing up for the following reason. All queens are shipped with attendant worker bees that feed and groom the queen while in transit. The attendants help to eat through the candy in the queen cage and when a hole is created that is large enough, the queen and attendants crawl through to join the hive. Sometimes one or more attendants may die while in transit or after the queen cage is placed within the hive. If the candy end of the queen cage is facing down and one or more attendants die, they fall into the candy and block the escape route for release of the queen. For this reason, always put the candy end of the queen cage facing up. If any attendants die they will fall away from the candy end and do not block the queen's release.

QUEEN WITH ATTENDANTS

REQUEENING IN THE FALL

Question - I have been advised to requeen my hives in the fall. Wouldn't it be better to wait until the following spring so the queen would be fresh and new for the coming season?

Answer - Many beekeepers can benefit from fall requeening. The colonies have come through a spring and summer build-up and many queens will be old and worn out. A good practice is to tag the nonproductive colonies during the flow. Such colonies should obviously be requeened. The proper timing is essential—and this time will vary from one location to another. As the fall season progresses, the brood rearing of the colony starts to decline. When brood rearing reaches a low level, it is an ideal time to requeen. It is a matter of balance between the queen being removed from the colony and the new queen being introduced—neither is at the peak of egg laying and the worker bees are more apt to accept the new queen. The old queen is more easily found as egg lying slows and the colony cluster reduces in size. When you open the hive, start your examination for the old queen by removing one side comb and then splitting the cluster in the middle - working both directions. The old queen is killed and the new one introduced, usually using the mailing cage for introduction. As the colony is put back together, sprinkle a little sugar syrup over the top bars of the brood nest.

Colonies requeened in the fall of the year will pay big dividends for the beekeeper the following spring and summer. In the spring, such queens perform just as a new queen—building up large colonies capable of maximum honey production.

INTRODUCTION OF QUEENS

Question - I've read that when you want to introduce a queen and her attendants that you have received in the mail *immediately* to your colony that you can spray both groups with a sugar, vanilla and water solution. This will keep both groups of bees more docile while hiving. My question is, if all the bees are sprayed with the above mixture, why are the additional instructions given to *destroy all* the attendant bees, even though they have been sprayed also. It seems to me that all the bees will go through the same cleansing and introduction process and it shouldn't make any difference.

Dan Perdue
Lincoln, NE

Answer - Without getting into a long-winded explanation as to why you are absolutely right, let me say you are. In the process above, you don't have to kill the attendants.

SWARMING AND ABSCONDING

Question - I'm confused. It seems as though some beekeepers I know use the words "swarm" and "abscond" to mean the same thing. Do they?

Dave Coovert
La Canada, CA

Answer - *Not really, but the meanings may be close enough that some confusion or misinterpretation may result.*

Swarming is a method, if you will, of reproduction. Remember back in high school biology, when they showed you how one bacteria would divide forming two complete and viable bacteria. Well that's kind of what happens to a colony of honey bees. For a variety of reasons when a colony of honey bees reaches a certain size a new queen is raised and the colony divides. One-half of the colony is headed by the original queen and one-half by the new queen. Generally the half with the old queen leaves (swarms) to find a new home. The other half headed by the new queen stays behind occupying the original hive.

Absconding is a method of colony survival. In some species of honey bees especially "Africanized" honey bees, the whole colony is genetically programmed to leave when nectar and pollen services dwindle. In Africa this is a survival technique that has been developed by honey bees over thousands, perhaps millions of years, to cope with sporadic nectar and pollen sources. Instead of staying in the hive when there is no nectar or pollen available, these honey bees leave (abscond) as a single colony unit and look for a better area to forage in. They may do this several times a year, depending on food sources available.

Question - I've been thinking about making several baby nucs to raise a few queens this spring. I'm planning on making them half the and width of a 6 5/8" depth super. My question is: What kind of entrance is used on these baby nucs?

Jess Juneau
Hessmer, LA

Answer - *Because the population is small in a baby nuc, the entrance should be small - generally, only big enough for one or two bees to pass. This small entrance is more easily defended and discourages robbing.*

Another important consideration is making the baby nuc entrance as distinctive as possible so that the returning mated queen can find her own colony. Don't put entrances together, stagger them. Use different colors on the entrances, paint a unique geometric shape on the facing wall on the nuc. Move the nucs around so that all entrances are pointing in a different direction and the trees and bushes used as landmarks aren't all the same. Anything you can do to indicate to the virgin leaving that this is home and can easily be found again is a plus. Remember all your efforts to raise a queen can be wasted if she can't find her own baby nuc.

DOUBLE GRAFTING

Question - Some queen producer ads in the *American Bee Journal* advertise "double grafted queen". I know what grafting is, but what is double grafting?

Dan Hendricks
Mercer Island, WA

Answer - *As you know, when queen producers graft the proper age larva into queen cell cups and these are, in turn, placed into a queenless colony, those larva will normally start being fed large amounts of special food called royal jelly by the colony's workers. This will determine whether the female larvae will develop into fertile queens or infertile workers. The feeling has been that when the larva is grafted into the cell cup that it is without food until the workers in that queenless colony start the feeding process and this short period of time without food is detrimental to producing the best quality queens. Someone, whose name has been lost to history, decided that if after the first day after grafting you removed the original larva and re-grafted another very young larva into the royal jelly pool deposited for the first larva, then feeding by the second larva could begin immediately and a higher quality queen would be produced.*

The workers will continue feeding the larva if the first larva is removed after the first day of grafting. An interesting fact is that the royal jelly being fed to the developing larvae that the colony is feeding for queens changes in composition after the first day of grafting. So, if you removed the original larva after the second or third day after grafting and then re-grafted a young larva into it, you would produce a much inferior queen than if you had left the original larva in. The food fed developing larva is very time specific. The nutrient composition of the feed fed developing larvae changes over time.

QUEEN FINDING AND MAKING SPLITS

Question - I always have trouble finding the queen. Could you tell me how to split a colony efficiently and not worry about where the queen is initially, but be able to find her later?

G. Bruns
Hertogenbosch, Holland

Answer - *If you make an effort to equalize the two colonies that you are splitting from one in regard to brood, honey and pollen, then knowing the location of the queen is fairly easy. One to two days after making your splits, open and examine your new splits. The split without the queen will now have queen cells. The workers have realized the absence of a queen and are trying to make one. The split with the queen will not have queen cells, but will have newly laid eggs in the brood cells.*

Question - I am brand new to beekeeping and am having a lot of trouble with bad queens. The first one was laying a few eggs here and there for a while and then became a drone layer. The second one was a double hybrid, but although I waited two weeks, she didn't lay any eggs. The current queen is laying eggs, but I don't know if there are as many as there are supposed to be. Any idea what might be wrong?

Mike Fleunweurth
Indianapolis, IN

Answer - *It is always difficult diagnosing these kinds of problems because there are so many variables that can influence the kind of problem you are having. On the surface the easiest thing to do would be to blame the queen breeder or breeders who supplied you with the queens. Any queen which does not lay or turns into a drone layer should be replaced by the breeder. Sometimes the virgin queen does not mate or mate completely and is thus incapable of laying fertilized eggs.*

Another possibility is she was injured during the trip through the mail to you. Perhaps injury occurred to her during her release and introduction. And then another thing comes to mind - were there sufficient workers to enable the queen to lay many eggs. Without a large supply of workers to feed the developing larvae, the queen will not lay. Many other things can affect the queen's ability to lay.

These are at least some of the things that may have happened to your queen.

Hang in there. The last queen you mentioned sounds as though she is laying well. Be sure that you continue feeding sugar syrup as a stimulative to egg laying until your colony is built up well enough to fill at least one whole brood chamber.

Question - I would like a real simple explanation of what supersedure is. I don't understand some of the explanations in the books.

Carl Karloske
Vesper, WI

Answer - *My dictionary gives the explanation of the word supersedure as "to take the place of."*

That's exactly what supersedure is in a hive of honey bees. It is the replacement of an old inferior queen by a young vigorous queen raised by the honey bees in the colony.

Various things happen to a queen as she ages that tell the colony that she is not as vigorous as she once was and needs to be replaced. The bees in the colony then raise replacement queens. When a new queen is raised, she takes the place of the old queen who dies or is killed.

IDENTIFYING THE QUEEN

Question - I'm in my first year of beekeeping. The bees that I purchased were from a neighborhood beekeeper. In my ignorance I bought a feeble queen and thus a weak colony. After expressing my discontent with the hive's stamina, the beekeeper located a swarm for me to catch. After I introduced my new swarm into the hive the two queens must have fought as I had two different races of bees in my hive, and all the younger bees were the same as the new swarm. I must admit that I have a hard time differentiating between the queen, drones and workers. So, my question is, can I introduce a new queen with packaged bees (maybe one pound or so) with an existing swarm? If so, will the two queens fight and the best one win? Or will they swarm? Or, should I take a beekeeping course?

Mike Clark
Hampton, VA

Answer - You can introduce a new queen to your colony some time in early spring. It's important that you learn to find and recognize the different kinds of bees in your hive so that you will become a better beekeeper. If a beekeeping course is offered in your area, take it. If not, call your county extension agent to find a knowledgeable beekeeper in your area who can take some time with you to show you the workings of the hive. To get back to your question, it's important to find the old queen in your hive and remove her before introducing a new queen. If you don't, then chances of your new queen being killed by the colony are about 100%.

So, between now and spring find a knowledgeable beekeeper to help you, and go to your library and check out or buy First Lessons in Beekeeping published by Dadants and read it cover to cover. If you do these things, when it comes time for you to order your new queen, your chances of success will be much improved. We also suggest that you become a member of your local beekeeping association. Good Luck!

Question - In one of my hives with a Carniolan queen and Carniolan workers, about half of the drones are black as one would expect. The remainder of the drones, however, are a yellow color like an Italian worker. They appear normal otherwise. Since drones develop from unfertilized eggs, I'm at a loss to explain the lack of uniformity in color. Can you help?

In addition, I'd like to be able to move drones from my outyard to where I'll be raising some queens. Can drones be successfully introduced into other hives?

Mark Ware
Elbert, CO

Answer - The bottomline answer is that your Carniolan queen is probably not a pure Carniolan queen and probably did not mate with only Carniolan drones. Virgin queens will fly up to 10 to 15 miles to drone congregation areas where drones from an additional several mile radius rendezvous. It's almost impossible to flood an area this size with only one race of drones. You will get drones of different races and mongrel combinations from other beekeepers' colonies and feral populations in these drone congregation areas. The virgin queen will then mate with 10 to 20 of these drones. Your chances of her mating only with the drones that you would like her to are slim. That's why you'll see drones and workers at times in the same colony that have different colors and temperaments.

So, your queen's mom and dad may not be of the same race. Also, your queen probably mated with Italian drones, as well as Carniolan drones.

Generally, drones will be accepted at least temporarily into other colonies. They drift successfully between and among colonies well. Whether or not you can artificially introduce them into other colonies, I really don't know. Let me know how it works out.

Question - I am a beginner and have three colonies of bees which I plan to keep as an avocation. I purchased these colonies early this spring as packages and each has a hybrid queen. Two of these colonies seem to be doing quite well during this honeyflow period. However, the third colony seems to be a "lazy" colony, and is considerably behind the other two colonies in their production. I am wondering whether it would be advisable at this point to replace the queen this fall and supplement the hive with another pound or two of bees early next spring?

Peggy Miller
Basco, IL

Answer - Requeen the lazy colony as soon as possible, even though there are probably other factors which are influencing this colony's production. It just isn't strong enough. We doubt that package bees will be necessary if you get the colony headed by a young queen and strong enough to winter by fall. Requeen and feed as necessary to go into the winter with adequate stores.

Question - I have often wondered if the following would work when requeening a hive? Prior to receiving your new queen, locate the queen in the colony to be requeened and put her in an old queen cage and then return her to the colony (inside the queen cage). When your new queen arrives, remove the old one and put the new one into the hive. By doing this, you would eliminate all the stress of locating the old queen and the transition would be smooth, eliminating the searching, frame by frame, for the old queen. We all know how that can be. Do you think that this would work out O.K.?

Tim Armstrong
Ashtabula, OH

Answer - Thanks for your question about queen caging and introduction. I don't think it would work though. Here's why.

The colony knows if they have a queen or how the queen they have is doing by chemical signals passed through the colony. Workers lick the queen's body and pick up chemical pheromones. These contact pheromones are then passed on to other workers by trophallaxis. Trophallaxis is the action of food/pheromone sharing by the workers when they pass these substances by mouth parts back and forth and among each other. Visualize a worker coming up to the screen of the queen cage and the queen allowing the worker to lick her. The worker then moves and shares this pheromone with another worker who, in turn, shares with another etc., etc., etc. The queen chemical in this case is passed through the hive and tells the colony that they have a queen. This is why the old queen needs to be removed a minimum of eight hours before a new queen is introduced because that's how long it takes for the queen substance passed around to disappear. The workers then begin preparation to rear a new queen or accept a new one. This is the time when queen introduction and acceptance is the greatest.

INFLUENCE OF QUEEN IN THE FALL

Question - I always assumed that the queen slowed or stopped egg-laying at the close of the honeyflow. Now I am told that an ideal queen is one whose egg-laying rate is high and continues well into late fall. Isn't this continued egg-laying and brood-rearing going to be using up valuable winter stores of honey and pollen before cold weather really sets in?

Answer - Producing brood in the fall when the honeyflow has ended does sound counter to good beekeeping practices until we remember that the colony needs young bees to go into the winter cluster. This is the time of year when the influence of the queen is greatly felt. Some queens continue their egg laying until late in fall, even after the beginning of the cold weather. There is also a marked difference in the rate at which queens lay their eggs. Some queens reduce egg laying at a much quicker rate than others, although there may be some brood in the colonies throughout the entire fall period.

The ideal queen is one whose egg-laying rate is high and continues well into late fall, thus furnishing a populous cluster of young bees for winter. The stronger the colonies are with respect to a population of young bees before winter, the better they will be able to survive the winter period. A large colony of relatively young bees, well supplied with stores of honey and pollen, also will be able to rear brood in the latter part of winter and early spring. Such colonies will usually come through the following spring with little spring dwindling and, with proper management of their steadily increasing populations, can be brought to the honeyflow in optimum condition for gathering a maximum crop.

Question - My dad kept bees for over 70 years near Upsala, MN to supplement his farm income and was a reader of ABJ for almost as long. I have kept up the tradition on a smaller scale to furnish honey to some of his old customers, my friends and relatives, but most of all because of my fascination of working with bees.

I would like to know if there is a simple way to raise one or two queens from a good-producing, problem-resistant hive. Even if I did not get honey from this divided colony the first year, I would like to try to build up bees native to the area.

Richard Anderson
Minneapolis, MN

Answer - I am glad that you are keeping up the family tradition of keeping honey bees. They are fascinating. Raising a new queen can be rather simple. To increase your colony numbers with new queens, here are some possibilities.

(1) In the spring if your overwintered colonies are strong, they may prepare to swarm. If you are conscientious in checking your colonies for swarming, you will notice the sure swarming sign-construction of queen cells. Once you see these, then you can divide your colony, giving it one or two cells and destroying the rest. Or divide, kill your old queen and give each divide frames with cells in order to requeen and start a new colony all at once.

(2) Divide your colony, leaving the queen in one divide and in the other divide be sure that there are eggs and/or newly hatched larvae in one or two frames. The divide without the queen will start raising a new queen, using the young larvae,

There are other more involved procedures that can be used to raise excellent queens, but for just a few try the above. They'll work for you.

Question - Can laying workers be distinguished by their physical appearance? Not for the first time, today I saw a bee which made me wonder. At first I wondered if she was a virgin queen. Her abdomen was only a little longer, but enough to catch my eye. She was exhibiting laying behavior, peering into cells and thrusting her abdomen a little into cells. Distinguishing, though, was the fact that her thorax was not bald as is a queen's.

The hive still has a marked laying queen that has been laying a complete pattern for 6 weeks. There have been not only no queen cells during this period, but no drone cells either. Prior to this queen becoming established, the hive had been queenless for three weeks owing to a series of misadventures during the requeening.

Dan Hendricks
Mercer Island, WA

Answer - The queen produces a pheromone which suppresses the development of the normal functional ovaries in the female worker honey bee. When there is a drop in the amount of this pheromone because of preparation for swarming or the queen is just not producing enough, then laying workers develop. The laying worker only changes in regards to ovary development. I suppose that with larger ovaries that the abdomen of the worker would become somewhat larger also, but I couldn't find where any researchers had made those measurements. This laying worker will do all the right activities. Check the cell, clean it, measure it, etc. but her abdomen will not reach to the bottom of the cell and she cannot completely control the quantity of eggs released. So, you get the pattern of multiple eggs adhering to the sides of cells. Because the eggs are unfertilized, if they do survive, it is as small drones.

Anyway, the interesting info I found is this:

During the swarming season workers with developed ovaries are found in 20% to 70% of colonies. One researcher, Koptev, in 1957 found that in otherwise normally appearing colonies that laying workers were found year round at a 7% to 45% level.

It sounds as though laying workers are a somewhat normal event in a large percentage of colonies instead of an abnormal condition that we generally consider.

Question - I have been advised to requeen my hives in the fall. Wouldn't it be better to wait until the following spring so the queen would be fresh and new for the coming season?

Steve McDaniel
Elizabeth City, NC

Answer - Many beekeepers can benefit from fall requeening. The colonies have come through a spring and summer build-up and many queens will be old and worn out. A good practice is to tag the non-productive colonies during the flow. Such colonies should obviously be requeened. The proper timing is essential - and this time will vary from one location to another. As the fall season progresses, the brood rearing of the colony starts to decline. When brood rearing reaches a low level, it is an ideal time to requeen. It is a matter of balance between the queen being removed from the colony and the new queen being introduced; either is at the peak of egg laying and worker bees are more apt to accept the new queen. The old queen is more easily found as egg laying slows and the colony cluster reduces in size. When you open the hive, start your examination for the old queen by removing one side comb and then splitting the cluster in the middle - working both directions. The old queen is killed and the new one introduced, usually using the mailing cage for intro-duction. As the colony is put back together, sprinkle a little sugar syrup over the top bars of the brood nest.

Colonies requeened in the fall of the year will pay big dividends for the bee-keeper the following spring and summer. In the spring, such queens perform just as a new queen - building up large colonies capable of maximum honey production.

Question - I am planning on starting a new hive this spring. I now have a colony of Italians which are doing fine. I want to start a colony of Caucasians. Could this cre-ate a problem for me in cross breeding? Should I requeen more often in a situation like this?

Craig Pullman
Arma, KS

Answer - Starting a colony of Caucasian honey bees next to a colony of Italian honey bees will not cause any problems. If one of those colonies should supersede the existing queen or issue swarms, then the replacement queen or swarm queen will, of course, mate with as many as 10-20 drones from a several mile radius. This means that any drone from your colonies, your neighboring beekeeper's colonies, or any feral colonies has an opportunity to mate with one of these virgin queens and affect the progeny in them. If you have marked queens, it will be easier for you to keep track of the current queen in your colony.

Question - I am having a problem that I hope you can help me with.

I have been a beekeeper for over 20 years and have never had a problem with queen introduction. However, this year I have had at least 5 queens that have failed to be accepted. I have used various methods. This year I have used the mailing cage and the push-in cage as described in an *ABJ* article on queen introduction. I have also tried intro-duction using a nucleus (5 frames of bees with brood taken from one of my hives).

When using the mailing cages, the queens disappeared as soon as they were released.

When using the cage pushed directly into the comb, the bees tunneled under the wax and killed the queen.

When introducing using a nucleus, the bees insisted on raising queen cells. I kept destroying the queen cells and the bees kept building new cells until there was no more brood to hatch. After all brood was hatched, they still would not accept the queen. Again, using a nucleus, I shook all of the bees through a queen excluder to insure there were no unfound queens. There were no laying workers and no brood that could be used to build queen cells. Everything should have been ideal for introduction. However, after 7 days when I opened the hive, the bees were balling the queen.

I am not the only beekeeper in my area who is having this problem. A neighbor who has been keeping bees for over 30 years now has 3 hives that are queenless and without brood.

What help can you give us (A procedure would be nice) that would allow us to successfully introduce queens. We are in dire straits and in danger of losing our hives,

We have had a very difficult season this year in New Jersey. Many days have been over 100 degrees and we have had no rain for weeks on end. I don't know if this is a factor or not.

Lowell Barrows
Hoboken, NJ

Answer - *It sounds as though you have done just about everything possible for successful introduction of a queen. I only have two additional thoughts for you.*

(1) Are you sure whoever was supplying you with queens was supplying mated queens? By your description, it almost seems as if there was another queen or laying workers around because they sure didn't recognize the new queen as a queen. If she wasn't mated, she might not be producing the pheromones necessary for acceptance.

(2) You mentioned temperature (over 100°F) and drought. When I think of these conditions, I think of lack of nectar-producing plants. Many times it is much more difficult to requeen with high temps and little or no nectar coming in. I always put some kind of a feeder with sugar or syrup on colonies to requeen in hot dry weather. Right off hand, I do not know what other reasons there might be for your problems.

Readers, this is not time to be bashful. Does anyone have any additional insight that I'm missing?

Question - What are the relative influences of the queen and the drone on characteristics of color and noncolor traits expressed by their offspring? Which characteristics are primarily determined by the queen, which by the drone and which equally? I've never seen this discussed.

Dan Hendricks
Mercer Island, WA

Answer - I've relied on Dr. Rob Page from the University of California in Davis for the answer to this question as he is vastly more qualified to give you an answer. "The queen and drone father contribute equally to the genetic composition of their worker offspring. However, workers will often resemble one of the parents, but not the other. Cuticular coloration is a good example. The late Will Roberts (USDA-ARS) demonstrated in his Ph.D work at the University of Wisconsin, Madison, that yellow color is dominant over black. Therefore, a black queen mated to a yellow drone will produce yellow workers. We have verified this result in my lab. The cordovan color trait is recessive. That means that workers must inherit cordovan genes from both the mother and father in order to have cordovan cuticle. A cordovan queen mated to a wild-type drone will not produce any cordovan workers. Other traits shown to have dominant or recessive inheritance are: hygienic behavior, many eye-color mutations, and defensive behavior. Ernesto Guzman, Department of Entomology, University of California, Davis, for his Ph.D. project, crossed highly defensive Africanized queens to gentle European drones and European queens to Africanized drones. The resulting hybrid colonies were very defensive like the Africanized parents. This demonstrates "nonadditive" inheritance even though both parents contributed equally to the genetic makeup of the colony."

Question - I have been in the beekeeping industry for going on 17 years. In my 17 years, I've never encountered this situation before, and was wondering if you, or any other beekeeper, could answer my question.

In Texas, on 3/6/93, I was out in a yard treating the hives and feeding the hives also. As I opened a particular hive a ping pong ball sized ball of bees, in the top corner of the hive, caught my attention. As I proceeded to smoke and dispense the small cluster of bees, I came upon a queen with an egg protruding from her abdomen. The bees were in the process of balling her. I tried to get her to go down into the hive by means to smoke, but to no avail. Then I tried smoking her and using my hive tool, very gently to get her to go down into the frames, again to no avail. She then attempted to fly and all she did for the first three attempts was to skim over the top bars. Then, on the fourth try, she successfully took off and flew. I stood and watched her for a few seconds until I lost her to the other bee flight.

My question(s) is this: I have never known a mated queen, after she has started laying to venture outside her hive, except for the purpose of swarming. So if that queen did not belong to that hive, how did she find her way into that particular hive and how is it possible for a mated queen to fly, especially with an egg protruding from her abdomen?

Bill Jensen
Cathay, ND

Answer - *I've never heard of the situation you encountered in a normal colony of European honey bees. One thing I have learned though is that beekeeping is many times more art than science and honey bees are capable of just about anything.*

A couple things flashed in my mind when I read your letter. One of these is kind of scary, so I'll leave it until last.

Sometimes with a newly mated queen that was produced for swarming reasons, she will be balled if the beekeeper manipulates that colony before swarming. Any new queen is in jeopardy of being rejected until she is fully producing pheromones that are accepted by the colony. You see this sometimes with introduced queens, rarely, and it may be assumed that it could happen with either supersedure or swarm-produced queens.

The second thought that flashed through my mind is the one that bothers me and that I am probably wrong about, but here goes.

You mentioned that it almost looked like the queen didn't belong in that colony. The reports that I have read about Africanized honey bees indicates that they can enter colonies that are newly queenless or preparing to swarm. The Africanized colony can "smell" the reduced pheromones from colonies in queen transition and many times reasonable and safely enter the colony with their queen and set up housekeeping.

Where you are in Carthage, Texas is quite a distance from where the line of Africanized bee penetration is so I am probably wrong, but it wouldn't hurt to have the USDA lab is Weslaco check some of your bees if you think this may be a possibility.

Question - I am a novice beekeeper. I started my first two hives early this spring from package Italians. Both colonies were growing in population at a good rate. In mid August I noticed one of the hives had a few bees that were grey with black bands mixed in with the orange with black banded Italians. At first I thought that they were robber bees. With closer observation I saw that they were bringing pollen into the hive. Today, September 11, I noticed that the better half of the bees are now grey and black. My guess is that somehow I lost my original queen and the colony reared a new one. I enclose a sample from the observed hive, two of each for comparison. Could you tell me what kind of mixed breed I have and if its one worth keeping?

Todd Hodgson
Marysville, WA

Answer - *Breeding of virgin queens in somewhat different from breeding other more well known animals. It is easy to know who is the father of a calf when there is one bull servicing some heifers. Virgin queen honey bees will mate with up to 10 to 12 different drones. They can store this sperm in a balloon shaped structure within them for several years. This arrangement of multiple matings makes it difficult for the queen breeder to 100% guarantee that his specific drones will always mate with his virgin queens. Try as they might to control who mates with whom, it isn't always possible.*

Another interesting thing is that in the sperm storage balloon, called a sper-matheca, the sperm from one drone does not mix with the sperm deposited by other drones. The sperm is stored in layers in this structure. So, you may see the workers at one time all looking the same for a few weeks, then notice the workers' color changing as one drone's sperm is used up and another begins being used. You may still have the original queen in your colony, but she may have mated with different races of drones. Or, as you have noted, the original queen was lost or superseded and she was mated with a mixture of drones in the area. Still another possibility is that the queen breeder sent you an Italian package of bees with a Carniolan or Caucasian queen. As the old Italian workers died, the eggs laid by the dark queen developed and began emerging to take their place.

This is not necessarily a bad thing. You now have an opportunity to see if the queen you have passes on the quality you desire in the workers. Is she worth keeping? Only a season under her belt will tell. If you want to requeen with an advertised particular race, you have that option also. It all depends on how much of an experimenter/researcher you are.

Question - I plan to raise about 20 queens this spring to put in four-frame nucs. This will be done in April which is not too cold in this area, but not really warm, ranging from the high 60s to the 80s F., depending on the day. The queen cells will be produced using the Jenter method so they will be on a cell base.

I want to know what is the best method to transport the cells about 100 miles to be placed in the area where the nucs are to be made. They will be placed in the nucs two days before they hatch.

1. How warm do they have to be kept?
2. Can they be laid on their side?
3. Should I leave them on the cell bar and move them with bees covering them?
Please give me any information that might be helpful.

James Neagle
Richmond, VA

Answer - *I am glad that you are planning on raising some queens this spring. I'm confident that you will be able to raise quality queens, learn a lot while you are doing it and save some money.*

It sounds as though you have done your homework on what to do and what you need to do. Because of this homework, you obviously know that you will get better acceptance using these cells than mated queens. It's a good idea.

The easiest and safest way to transport the cells is to leave them on the cell bar in a queenless nuc with plenty of bees. The bees will then be able to control the temperature and humidity to the optimum for transportation. This would be 93ºF at 50% relative humidity.

The next safest way is to adapt a small Styrofoam picnic type cooler to hold your cell bar. Using warm (93ºF) moist towels, you can approximate hive conditions somewhat and reduce a great swing in temperature while transporting

*Even though these virgin queens in their cells are very far along in develop-
ment and the trip probably wouldn't damage their wings if placed on their sides and
bumped along on your trip, I would try to keep them in their cell bar in the proper
position as long as possible.*

*Other than these few conditions which are really rather easy to maintain, the
whole procedure is rather painless.*

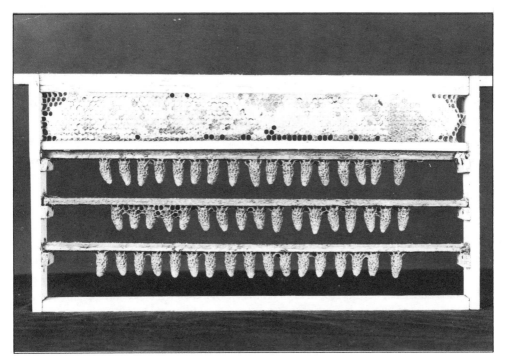

QUEEN CELL BAR

Question - I have a 2-frame observation hive inside my house with a flight tube
through the wall. The colony had a lot of chalkbrood and was not developing very
well, so I decided to requeen it using a queen which had shown little or no sign of
chalkbrood. Because it is a lot of trouble opening up the hive - I have to plug the
wall entrance, carry the hive outside and unscrew the front glass panel before I can
even start to take out the combs - I decided to try a quick switch queen introduction.
I first filled a household spray bottle with diluted honey, then opened up the hive,
found and killed the old queen. Then I liberally sprayed all the bees on both sides of
each comb with the diluted honey. I put the new queen right on the comb, gave her
a spray of honey as well, and closed up the hive as fast as possible. I thought I
should leave the shutters on for a couple of days, but my wife was keen to see if the
new queen bee had accepted, so around 5 p.m. we took a quick look. What we saw
was apparently the queen being balled. The bees were tightly packed in a knot like

this for 2 1/2 days though the knot moved around the frame somewhat. On the third day the queen was walking about and laying normally, and now some weeks later the hive is very strong with no sign of chalkbrood.

Is this normal behavior of bees towards a new queen, but we don't usually see it? How often do beekeepers requeen observation hives?

Dave Dawson
Manitoba, Canada

Answer - Balling of a queen usually results in damage or death to the queen. I'm surprised your queen survived. But she did and is laying so that is a good sign. As you know, honey bees have a highly developed system of chemical signals (pheromones) which help organize the hive and enhance its survivability. Part of honey bee survival relies on the colony being able to defend itself from intruders. These intruders are many times recognized by the fact that they do not possess the same chemicals (pheromones) as the rest of the honey bees in the colony.

You introduced an intruder into the colony. The bees did what they could to isolate the intruder and most likely pulled at her wings, legs, antenna, etc. and tried to sting her. This is a common reaction to perceived danger from a non-colony member. If you had taken more time to artificially isolate the queen in her own queen cage until the house bees were able to collect her chemical pheromones and distribute them throughout the hive, it would have protected her more. Then when you released her, no balling would have occurred.

The next time you requeen your observation hive, just to protect your queen and reduce the chances of her being killed, go a little slower. You were lucky this time.

Question - Is a double hybrid queen better, equal or inferior to any other queens advertised for sale?

Mike Fillenwarth
Indianapolis, IN

Answer - Honey bee queens can sometimes display vastly variable characteristics from their mother. The reason for this is because of how virgin queens mate. In an open mating plan, as used by most queen breeders, virgin queens that have been produced are allowed to fly and mate in the open. The virgins will mate with upwards of a dozen or more drones. The queen breeders try to flood the area with drones they feel will contribute positively to the characteristics of the progeny produced from the queen. These virgin queens will fly up to 10 to 15 miles in search of a drone congregation area in order to mate. As a result of this distance, it is almost impossible to have only the dozen or so drones be the ones the queen breeder would like to have mate with this virgin queen. So, there is a mixture of drones from the queen breeder, feral colonies, other beekeepers, etc., that contribute their genetic code to the workers that will be produced. This is why in many colonies some workers are yellow and then a group of darker workers are produced and then maybe back to yellow and so forth. These workers also display different degrees of foraging capability. Some will collect more nectar, pollen or propolis. Some will be more defensive than others or forage earlier or later than their sisters.

All of that points out that there is great variability among queens and their offspring when the queens are open-mated. The answer to your question is, it doesn't really matter.

Classroom Comment -

Dear Mr. Hayes:

This letter is in response to the October issue Classroom article concerning the N.J. beekeepers problem with queen introduction.

There are many factors that can greatly reduce the successful introduction of queens and I would like to discuss a few of the most common, but least recognized.

1. Robbing - Any colony that is or recently was under the stress of robber bees is almost certain to reject any introduced queens. Sometimes robbing only lightly occurs, so slight that to the untrained eye it has the appearance of normal flight activity. The level is most acute to nucs and weakened colonies that still have defenses, but at a reduced level.

2. Candy - Candy not only provides substance to the queen and her accompaniment of attendants, but is also a timing mechanism. I have purchased queens through the mail and have noted upon their arrival that the candy was as much as 50% consumed. It would not be advisable to introduce with these cages as it would result in a premature release and lower the acceptance rate for the queen. I would also like to add that every now and then an article appears that promotes the puncturing of the release candy with a nail or other object. This practice can only cause you troubles as it shortens the release time to as little as 24 hours.

Cells - While it is a good idea to wait 24 hours before introducing a ripe queen cell to a colony after removing the existing queen, it is not sound practice to do so when introducing a mated queen. Once queenless, a colony will almost immediately start construction of cells. If a beekeeper waits too long to place new queen in the hive, advanced cell development is a danger. Even when a new caged queen is placed in the colony immediately, the bees will begin to build cells for there is the absence of a *laying queen*. If the queen is released on time, the bees will tear down these beginning cells. If, however, the cells are closed, they seldom do and the presence of a closed cell can cause reluctance of the bees to accept the new queen.

Marcus H. Kaiser
St. Jacob, IL

Marcus, thanks for your well-thought out answer. I'm sure it will help our readers.

Jerry

Question - What's the easiest and best way to start a two-queen colony?

Howard Twilley
Miami, FL

Answer - *Pages 633 through 635 of* The Hive and the Honey Bee, *Chap. 14 took the words right out of my mouth.*

"*The basic reference for two queen management systems is a 1976 USDA report by F. E. Moeller entitled* Two Queen System of Honey Bee Colony Management. *This system is also described in the USDA Handbook #335 entitled,* Beekeeping in the United States.

"*F. E. Moeller is quoted as saying, 'The establishment of a two-queen colony is based on the harmonious existence of two queens in a colony unit. Any system that ensures egg production of two queens in a single colony for about two months before the honey flow will boost honey production.' Moeller goes on the state that the population of such a two-queen unit may be twice that of a single queen unit and will produce honey more efficiently. He also claimed that the two-queen colony would enter the winter with increased pollen stores which encourages the production of a larger population of young bees in the spring and thus is an ideal unit for producing another two-queen colony the following year. Moeller's claims are the basis for most of the two-queen systems in use today.*

"*Using Moeller's work as a standard, the following general procedures can be followed in developing a two queen system.*

1.   *Start with strong overwintered colonies.*
2.   *Build the colonies to maximum strength in early spring by feeding pollen patties and sugar, if appropriate.*
3.   *Approximately two months before the start of the major flow the colony is divided and a new queen is introduced into the queenless portion. The old queen, most of the younger brood, and about 1/2 of the adult bees are placed in the bottom section of the hive. Either an inner cover with the hole plugged or a thin board is placed over the bottom unit. The new queen, most of the capped brood and the remaining bees are established above the division board.*
4.   *The upper unit is provided with an auger hole which serves as a hive entrance for those bees.*
5.   *Two weeks after the new queen's introduction the division board is replaced with a queen excluder.*
6.   *During the initial division the bottom unit was provided with an empty brood chamber so that the old queen would have plenty of room to lay eggs. After the new queen begins laying, another brood chamber is also added to the upper unit.*
7.   *Brood production may advance quite rapidly and it may be necessary to practice reversals (of the brood chambers) in both the upper and lower units. This may have to be accomplished every 8-10 days until about 4 weeks before the end of the major nectar flow.*

8. Honey supers are provided to the upper and lower units as needed for honey storage. Extracted supers are returned to the units for additional filling. Moeller cautions that a two-queen unit requires considerable storage space and when a one-queen colony might require one super, then a two-queen colony might require two or three empty supers.

9. The advantage of a two-queen system is lost about four weeks before the end of the major flow. Eggs being laid at this time will take at least four to five weeks before they become foragers and can collect honey. The queen excluder is removed and the units are combined. The result, according to Moeller, is usually the death of the older queen. This insures that the colony will normally go into the winter with a young queen.

"The two-queen system does seem to demonstrate some advantages under beekeeping conditions found in the northern states where there is a fairly long (about two months), intense and predictable honey flow. The advantages are not as obvious to southern beekeepers or any beekeeper where the honey flows are shorter, less intense and more unpredictable. A summary of the advantages and disadvantages of the system are provided for those who might be interested in experimenting with this production system.

Advantages:

1. The beekeeper tends to equalize the colonies during the setup process, insuring that all have adequate bees to take advantage of the coming nectar flow.
2. Swarming and supersedure tend to be reduced because of the regular introduction of new queens and the swarm prevention practiced in the management system.
3. The system produces colonies that should overwinter well because of the presence of young queens and good food stores.
4. Most beekeepers get good queen acceptance using this system.

Disadvantages:

1. The two-queen system may not be advantageous for southern beekeepers or beekeepers in areas that do not have long lasting, intense nectar flows with predictable start and stop times.
2. Significant labor and timing are necessary for a successful system.
3. Since the units may be very productive, they may become quite tall as honey supers are added and the risk of the hive tipping is of concern.

"The best advice to the beekeeper considering a two-queen system is to experiment with a few hives of bees and see how the system works for him/her."

Question - As an experiment, I am considering replacing a Midnite Queen with a Buckfast queen. Do you have any predictions on how she would be accepted?

Mark Ware
Colorado Springs, CO

Answer - If you make sure that you remove the Midnite queen a minimum of 8 hours before putting in the Buckfast Queen cage and let the bees release her on their own, this shouldn't be any problem.

Bees are very accepting of any queen as long as they are newly queenless.

QUEEN REARING

Question - I enjoyed reading your article on "The Hopkins Method of Queen Rearing". I would like to know where I might obtain more specific information and instructions on this specific method. It appears that this method is so simple and fundamental that it is ridiculous.

Thanks for any assistance you might give me.

James A. Cosgrove
Metairie, Louisiana

Answer - Thank you for your letter and interest in my article on the "Hopkins Method of Queen Rearing," that appeared in the May 1991 American Bee Journal.

You asked for specific information and instructions on this method of queen rearing. A copy of the section on the Hopkins Method from Frank C. Pellett's 1929 book, "Practical Queen Rearing" appears below. As you can see, the information is simple and direct. I know many feel it is too simple and easy and as a result do not even try this method. It is a shame that in our society we think that the more complicated something is, the better it is. Not so!

BIG BATCHES OF NATURAL
CELLS BY THE HOPKINS OR CASE
METHOD

Many extensive honey producers who desire to make short work of requeening an entire apiary, and who do not care to bother with mating boxes or other extra paraphernalia, make use of the Case method, which has been somewhat modified from its original form. This method is advocated by such well known beekeepers as Oscar Dines of New York and Henry Brenner of Texas. The plan was first used in Europe.

To begin with, a strong colony is made queenless to serve as a cell building colony. Then a frame of brood is removed from the center of the brood nest of the colony containing the breeding queen from whose progeny it is desired to rear the queens. In its place is given a tender new comb not previously used for brood rearing. At the end of four days this should be well filled with eggs and just hatching larvae. If the queen does not make use of this new comb at once, it should not be removed until four days from the time when she begins to lay in its cells. At that time nearly all the cells should be filled with eggs and some newly hatched larvae.

This new comb freshly filled is ideal for cell building purposes. The best side of the comb is used for the queen cells and is prepared by destroying two rows of worker cells and leaving one, beginning at the top of the frame. This is continued clear across the comb. We will now have rows of cells running lengthwise of the comb, but if used without further preparation the queen cells will be built in bunches that will be impossible to separate without injury to many of them. Accordingly, we begin at one end, and destroy two cells and leave one in each row, cutting them down to the midrib, but being careful not to cut through and spoil the opposite side. Some practice destroying three or four rows of cells, and leaving one to give more room between the finished queen cells.

We now have a series of individual worker cells over the entire surface of the comb, with a half inch or more of space between them. The practice varies somewhat with different beekeepers beyond this point. However, this prepared surface is laid flatwise with cells facing down, over the brood nest of the queenless colony, first taking care to make sure that any queen cells they may have started are destroyed. In general, it is recommended that the colony be queenless about seven days before giving this comb. By this time there will be no larvae left in the hive young enough for rearing queens, and the bees will be very anxious to restore normal conditions. Some beekeepers simply take away all unsealed brood, rather than leave the bees queenless so long.

As generally used, this method requires a special box or frame to hold the prepared comb. This is closed on one side to prevent the escape of heat upward and to hold the comb securely in place. Some kind of support is necessary to hold the comb far enough above the frames to leave plenty of room for drawing large queen cells. It is also advisable to cover the comb with a cloth which can be tucked snugly around it, to hold the heat of the cluster. By using the empty comb-honey super above the cluster, there is room enough for the prepared comb and also for plenty of cloth to make all snug and warm.

Strong colonies only should be used for this, as for any other method of queen rearing. If all conditions are favorable, the beekeeper will secure a maximum number of cells. From 75 to 100 fine cells are not unusual. By killing the old queens a day or two before the ripe cells are given, it is possible to requeen a whole apiary by this method with a minimum of labor. According to Miss Emma WIlson, it is possible to get very good results by this method, without mutilating the comb, although it is probable that a smaller number of queen cells will be secured. By laying the comb on its side as practiced in this connection, the cells can be removed with a very slight effort and with a minimum of danger.

Question - I know how worker bees induce the queen to lay unfertilized eggs by providing drone cells. But what is the situation which causes her to lay such eggs in a worker-sized cell? Is it the queen's initiative or the workers'? Do we know?

Dan Hendricks
Mercer Island, WA

Answer - If the queen has mated and she has plenty of sperm stored in her spermatheca, then I would be inclined to venture a guess that an unfertilized egg laid in a worker cell is simply a mistake. I'm not positive what the stimulus is that releases sperm as an egg travels by the spermatheca for fertilization, but I'm guessing that sometimes the sperm are released too early, too late or not at all. Just as our bodies do not function perfectly all the time, the queen's body I'm sure doesn't either. Couple this possible sperm release/egg situation with sperm or eggs which may be imperfect so that fertilization does not take place, and it's remarkable that the queen can control this activity at all.

Instrumental Insemination Device

Question - I live in Arizona and some of my colonies have just been identified as being Africanized. I have been told that instead of having to destroy them, I can just requeen the colonies. My question is, where can I get queens this early in the year so that I can requeen?

Gordon Shumway
Malmac, AZ

Answer - Even though we all have known for years that Africanized bees would be coming and would be a problem, my brain never made it to the spot where winter requeening in some parts of the country would become a reality. As you know, the bulk of the queen rearing industry is in the southern areas of the U.S. But even with warmer temperatures, most of the year, they still have a winter which is not optimum for raising queens. But if queens were available, there are many companies in many states - Florida, Arizona, South Texas, New Mexico, Southern California, and Hawaii - that come to mind. Beekeepers may need queens at this time of year and still could requeen successfully. The queen rearing industry will adjust and be able to make queens more readily available for these southern africanized areas in the future.

For your immediate problem though, you will have to go through this and past American Bee Journals and contact queen breeders that you recognize from their location as perhaps having queens available. I'm confident that you will find some.

Question - I am interested in trying the "Hopkins" method of queen rearing. I read about this method in the August 1991 Classroom column. I would like information and instructions on how to put the ripe queen cells into a queenless colony, when and where to put them, and how many. Thank you for any assistance you might give me.

John Williamson
N. Bennington, VT

Answer - *I'm glad to hear that you will be trying the "Hopkins" method of queen rearing this coming season.*

Any method which uses queen cells for requeening will need the same conditions. The most important condition is that the colony be queenless for at least 12-24 hours. Couple this with a fairly populous hive, developing brood and a nectar or sugar syrup source and you may successfully introduce a queen cell.

Carefully remove the cell from the face of the comb, make an indention with your hive tool or thumb in the face of the comb you will be putting in the colony to requeen and gently press the queen cell into it. Obviously, be careful and gentle while doing this as the developing queen may be damaged. If you insert the cell after the 13th day of development or so your chances of injuring her are greatly reduced.

Have a good time with this and good luck.

Question - In the January *American Bee Journal* the first question in "The Classroom" was a question by Dave Dawson from Manitoba, Canada on bee behavior around a newly installed queen a few hours after installation.

I feel there may be a better answer than the one given. When the bees were sprayed with diluted honey, a few bees accepted the queen as their queen. The other bees cleaned themselves and the hive, but had not yet accepted the queen and, therefore, were antagonistic to the intruder, the new queen. So, the few bees who had accepted her formed a protective barrier around her so that she would not be balled. If she had been balled, she would have died. One queen cannot protect herself against many antagonists.

I had a good queen with very few bees in an observation hive. I added enough bees so that three-fourths of them were new to the observation hive. The bees formed such a barrier around her that she was squeezed flat against the glass. After 48 hours, they released her. She was accepted, began her queenly duties, and they all lived happily together until when I put them in a hive for winter, again uniting them with a queenless hive. They made it through the winter in good shape.

Yes, both Dave and I were fortunate our queens had some protection by some bees. But I believe that both of us have observed the bee behavior that probably often takes place when two hives are united. But since very few of our hives have glass, we have failed to observe and document this protective behavior and thus have remained ignorant of it.

Paul Goossen
Homestead, IA

Answer - Thank you for your letter and observation. I have never heard or read about a small group of bees selectively protecting their queen, but it is entirely possible. As you know, nothing in our knowledge of a colony of honey bees is truly black or white. Bees being the adaptable insects they are, are certainly capable of some behaviors with which we are not totally familiar.

Question - I would appreciate you sending me *Ingredients* of queen candy and how to *properly mix it.*

Jesse Juneau
Hessmer, LA

Answer - According to Chapter 23 of The Hive and the Honey Bee, "Queen Cage Candy is made from powdered sugar kneaded into a firm, but not hard dough with high fructose corn syrup." It's not any more complicated than that!

COMMERCIAL QUEEN MATING YARD

CHAPTER 2

HONEY BEES WORKERS/DRONES

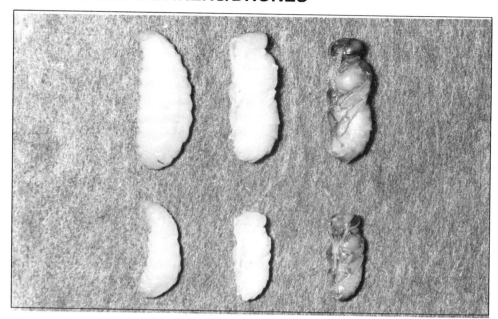

TOP - DRONE LARVA AND PUPA
BELOW - WORKER LARVA AND PUPA

Question - Bees, according to their age, develop certain tasks. Does this mean that when bees are foragers they can't go back to being nurses again? I would most appreciate knowing how rigid this behavior is in honey bees.

Andrew Rivera
Peru, South America

Answer - Sooner or later every creature that is rigidly fixed in its environment becomes extinct. Without some flexibility and adaptability, mental and socially, survival at some point, because of environmental changes, becomes extremely difficult.

Since honey bees have a fossel record dating back millions of years, they must possess this flexibility and adaptability necessary for survival. Specifically you asked about foragers becoming nurse bees again. Yes, they can and it happens all the time, especially in package bees in the U.S. Picture bees at random being collected and put into a small shoe box size cage with a new queen and shipped to some destination across the country. Those bees are of every age distribution imaginable and most undoubtedly will have to help establishing a new brood nest, taking care of their queen, warming and feeding the brood. The older foragers will revert to nurse bees. But they are not as good of nurse bees because the hypopharyngeal glands in their heads which manufacture the royal jelly/brood food have shrunk as foragers and never get their size back as large as when the bee was younger. The foragers turned nurse bees produce food, but not in the quantity of a younger bee.

Question - I was recently looking at an observation hive and noticed a worker bee running around the hive vibrating up and down on the backs of fellow bees. This behavior continued for some time and was repeated by the same bee. Is this behavior related to the round and wag-tail dances or is it a completely different phenomenon?

Ken Haller
Pensacola, FL

Answer - I think the "dance" you were seeing is the DVAV "Joy" Dance. This stands for Dorso-Ventral-Abdominal-Vibration. There is a good layman's description of it on page 299 in the new edition of The Hive and the Honey Bee.

Researchers have not yet agreed as to exactly why or what the dance does specifically only knowing that some minor behaviors are changed or redirected by it.

I must commend you on your powers of observation. Good job!

Question - Do bees have a clotting factor in their blood? If bees do not have a clotting factor, would not the varroa mite be bleeding their host to death by puncturing their bodies?

I have read somewhere that if you pulled a leg off of an insect, it would kill it because of no clotting factor.

Larry Keeney
Russellville, IN

Answer - I had always heard the same thing that you have. When a honey bee has a break in its exoskeleton, the blood leaks out continuously until it dies due to a lack of a blood clotting mechanism. I had always accepted this as "fact" and didn't explore any further. You, on the other hand, took the "fact" one step further by adding varroa mites to the equation.

As you can probably tell, I was just as ignorant of the answer as the next guy. But I did know who to ask. Dr. Justin Schmidt of USDA/ARS Lab. in Tucson took the time to share his expertise in the area. The answer is something I never knew or was taught... it's very interesting.

Here is Dr. Schmidt's Answer: "Actually, the question brings up a couple of interesting points. It also indicates that there is more to bee biology and physiology than at first comes to mind. I spent several years bleeding bees for biochemical analysis of their blood (hemolymph) proteins. Adult bees do indeed possess a clotting mechanism. If you puncture the membrane behind the head and collect some hemolymph in a capillary tube, it will quickly clot and you cannot get it out. How we prevented this was by adding a anticoagulant called glutathione to the collection tube, much as a physician adds heparin to bags to prevent human blood from clotting. Bees also have blood cells that are much like human white blood cells that function to fight intruders and infection as well as aid in plugging holes.

"When you look at other insects and bee larvae, the story gets even more interesting. It turns out that by insect standards adult bees are at best only average "clotters". For example, cockroaches are exceptionally good clotters, so much so that it is very difficult to get their hemolymph into a glutathione solution before it clots. And then honey bee larvae appear to have almost no clotting mechanism at all. We did not need to add glutathione to larval hemolymph and it could stand for long periods of time without clotting. If you look at the biology of the insects in these three examples, it makes sense. Cockroaches are the universal food of every-thing and are always at risk of getting injured. They also live in places where acci-dental injury seems possible. Adult bees are in the less risky situation, but do on occasion get injured. Larvae are in nice smooth, protected beeswax cells and essen-tially never experience accidental injury or predation, so they have less need for a clotting system.

"Back to the question of mites. The black "scarring" in the trachea from tracheal mites is the result of oxidation and clotting of blood. It is in a very real sense a defense mechanism of the bee's system in attempting to control the damage of the mite. Varroa feeds on both adults and larvae, but reproduces and does better on larvae. Could there be a physiological connection with the clotting? I do not know, but it seems worth pursuing. Certainly adult hemolymph could easily seal any wounds left by feeding varroa. If a larva is punctured, it also stops bleeding. It appears that the elastic integument just seals up the hole. Also, insects do not have much "blood pressure" as they do not pump blood through veins but rather through the open hemocoel, so pressure to bleed would be reduced. Eventually, the puncture would be sealed either by a slower clotting system or by cellular growth."

Question - I am an amateur beekeeper who started beekeeping just a few years ago, and I find it a really wonderful and fascinating hobby, spending literally hours watching and working among these marvelous little creatures.

On recently requeening my hives, I came across something I hadn't previ-ously experienced. One of my hives was queenless, and one of the combs, which was a very dark brown, had quite a lot of largish cells at the bottom of the frame, which I took to be drone cells. As I was requeening, I decided to cut out and discard these drone cells, and on doing so a white substance, like watery milk, squirted out. There was no smell, bad or otherwise. Can you explain this please?

Anthony Gomez
Arequipa, Peru

Answer - *You are absolutely right in saying that beekeeping is a fascinating hobby. A hive of honey bees is a remarkable creation. Individually the honey bee is just another "bug" but communally, they take on aspects almost mammalian. I'm amazed every time I think about it.*

I believe the watery milk colored material that you saw while cutting out drone cells was simply the contents of their underdeveloped pupal bodies. I've experienced the same thing when scraping out drone comb. The pupae about midway through their development are a very delicate structure. When disturbed by breaking into their cells, the delicate body is easily damaged releasing the watery white body fluids that you saw. There would be no smell or odor from healthy nondiseased brood.

Without seeing the cells, we could not tell you for sure that they were drone cells, but there is another possibility. The "largish cells at the bottom of the frame" may have been queen cells that the colony was rearing in order to make itself queen-right again. However, you solved that problem by requeening the colony yourself.

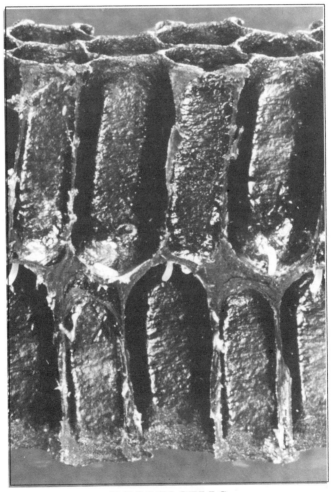

EGGS IN CELLS

ANCESTRY OF DRONES

Question - My grandfather used to really get me confused with the saying, "A drone never has a father, but always a grandfather." Would you explain this?

Answer - *The saying really is rather confusing until you know about the ancestry of drones. All female bees, whether queens or workers, have fathers because females are produced from fertilized eggs. The birth of a drone can come about in three different ways but remember that in each of the three cases the drone is going to be fatherless because drones develop only from unfertilized eggs. In each of the three cases, however, the drone will have a maternal grandfather because the unfertilized egg that produced the drone was laid by a female and all females have fathers.*

1) The queen bee, after mating with a drone, can lay two types of eggs, fertilized which produce workers or queens, and unfertilized eggs which produce only drones. The unfertilized egg received no sperm and hence, no father for the drone in this case of parthenogenesis, or virgin birth.

2) A virgin queen, one who has never mated with a drone, can lay only unfertilized eggs which, again, produce only fatherless drones.

3) In certain conditions such as a colony becoming hopelessly queenless or in other abnormal situations, a worker may lay eggs which will produce drones since the worker had never mated and the eggs were unfertile.

DRONES DYING WITHOUT CARE

Question - Is there any explainable reason why drones die within an hour after being placed in a vented jar?

Answer - *The drones apparently need continued care and feeding from the worker bees. Although drones will occasionally feed themselves from an open nectar cell, they solicit worker bees quite actively for care and tending. We have noticed the same thing ourselves here in our breeding program and that about an hour after the drones are removed from the hive, they start becoming extremely feeble. At that time, we place them back into the hive in a caged chamber, and let the workers revive the drones for us.*

DRONES IN THE WINTER CLUSTER

Question - Why do some colonies carry drones through the winter while others do not? Also is a drone reared from a laying worker egg capable of mating?

Answer - *Colonies carry drones through the winter only when they are queenless or have virgin queens. They would keep the drones for mating with the queen. Otherwise, they would drive the drones out of the entrance before winter to die from starvation or exposure. Yes, a drone reared from a laying worker egg would be capable of mating since he is a normal male.*

LAYING WORKERS

Question - Do you have any specific information on how to rid a hive of laying workers? The bees were shaken off the frames some distance from the hive but the laying workers must have made it back to the hive.

Answer - *There are a number of possibilities that can be followed with relation to getting rid of laying workers and requeening the colony. You do not say how far away you took the frames with the bees on them when you shook the bees off. Normally, the farther away you go, the more likely you will be to lose the laying workers. Quite often these are somewhat heavily developed in the abdominal area and, therefore, find it more difficult to fly back to the colony.*

If there are other colonies of bees, there are several things you can do using them. One would be to move the colony which has the laying workers, after having shaken the bees off the combs, to a new spot in the yard. Then, a number of combs of emerging brood could be taken from the other colonies and placed in the center of the hive. These combs should have the bees on them, but you must be careful not to move the queen from the colonies they come from. The push-in cage is the most foolproof method of requeening a colony under adverse conditions. We would suggest using this method and after shaking the bees off the comb, place the new queen on the comb and put the push-in cage over her within an area in which the young bees have partially emerged and are continuing to do so.

If the general age of the bees in the queenless or laying worker colony is getting too old for requeening, then it might be a good idea to unite this colony with a queen-right colony nearby. This can be done by using a sheet of newspaper over the top of the combs of the queen-right colony. Then, make small slits in the sheet of newspaper and place the hive body from the queenless unit on top. The bees will soon eat their way through and mingle together and make one strong colony of the two units. The bees from the now strong colony will dispose of the laying workers.

Question - I have kept bees for six or seven years as a hobby. Here is my problem. I hived a small swarm of bees that had settled low and it was easy to get them into the hive. I put them in a 8-frame brood chamber with six frames of foundation and two frames of young bees and honey which I had taken from the parent hive. Later I looked in the hive and couldn't find a queen. The foundation was partly drawn and had honey and eggs, and bees were bringing in pollen. About half the cells had eggs in them, two or more eggs in each cell. Also on three frames they had built queen cells up on the frames. These cups all had eggs in them, more than one in each cell and as many as twelve eggs in one cell cup. Now, do I have a laying worker? Could I reunite these bees back with the original colony they swarmed from? What would be the best thing to do with them?

Answer - *A queen will often lay more than one egg in a cell when she has only a few cells available and a young queen will also do this but it sounds as though you have a laying worker. Laying workers are a problem and requeening is difficult. Unless you have a fairly strong colony of young bees, it would be difficult to maintain them as a separate colony and we would suggest that you unite them with a queenright colony using the newspaper plan.*

LIFE OF A WORKER

Question - If the life of a worker bee is so short, how can they live all winter?

Answer - The life of a worker bee is usually shorter in the summer, depending on how much work they are doing – brood rearing, flying, or heavy foraging which soon wears out the wings. A bee that is in a cluster can live much longer than the summer bee and the young bees going into the winter cluster can survive until the young bees start coming on the following spring.

Question - How long does a worker live?

Answer - The length of life of the individual worker honey bee varies tremendously at different times of the year. Experiments with marked bees have shown that in a normal queenright colony in March the average expectation of life on emergence is about 35 days, but in June it has become reduced to an average of about 28 days, about 9 of which are spent foraging. On the other hand many of the workers that are reared during September and October live throughout the winter. There is a record of such individuals living for 304 days. The three factors would seem to be: (1) the amount of pollen consumption; (2) the amount of brood-rearing, and (3) the amount of foraging outside the colony.

BEAUTIFUL FRAME OF CAPPED WORKER CELLS

CHAPTER 3

COLONY MANAGEMENT

Question - I have a couple of questions which you may be able to clarify. I often read that beehives placed in a row should be painted different colors so that returning foragers may recognize their own hive, thereby reducing drifting to nearby ones. However, V. and C. Gould in their book "The Honeybee" Scientific American Library, page 196, write: "In fact there is evidence that, though they learn what their hive looks like ...cannot store its color (in their memory) even if it is brightly painted." The above statement is also contrary to what I read in *ABJ* Oct. 1992, P. 657 "Drifting Bees" by Steve Taber.

If the bees cannot recognize their hive by color, I am sure many other beekeepers would prefer to paint their hives like me to match the surroundings, camouflage color, to protect from vandalism.

The other question is: Is there a blue color pollen? In late September of this year I was sitting by a hive watching the comings and goings of its occupants. I saw a returning forager with both baskets full of electric blue pollen. I continued watching this hive for another half hour without seeing any other bee with blue pollen.

John Bunicci
Shoreham, N.Y.

Answer - *These are the things that researchers have found out about honey bee orientation and navigation. Honey bees use landmarks (shapes), polarized light, UV light, the position of the sun and moon and the magnetic fields of the earth as cues and clues to where they are and where they want to be. With this vast ability, you may wonder why a honey bee would ever get "lost" and drift to another colony. Honey bees can navigate and orient extremely well, but we have set up some abnormal conditions to which they must respond. Most apiary sites consist of more than one hive which is within several feet or less of another hive which looks and is colored the same as the first. Perhaps there are 10, 20, 30 or more hives in this area. This makes it tough for a creature without tremendous long-term memory ability. I'm impressed they even find their way back to within several feet of their own hive at all. To make it easier for all bees to orient to their own hive, the hives should be broadly scattered with excellent visual clues available for each individual hive – a large tree for this one, a large bush for another, etc. You can also provide visual cues for them with large geometric shapes attached to individual hives. These actions will help your bees find their own hive.*

I have never seen electric blue pollen in late September, although there are several plants which yield blue pollen in either the spring or summer. According to Dorothy Hodges' book, The Pollen Loads of the Honey Bee, two of the blue-pollen species yield their pollen in summer – Queen Anne's Thimble (Gilia capitata) and Phacelia tenacetifolia. The other sources such as Siberian squill yield their blue pollen in the spring. There is also the chance that the single honey bee had collected a nonpollen source that had a blue color.

THE PACKAGE BEES HAVE ARRIVED

Question - I am new to beekeeping and enjoy your column "The Classroom" since many of my questions are asked by others.

The one question that I have not seen asked is in regards to an upper entrance. Even the Dadant and Sons book *The Hive and the Honey Bee* has not been very helpful.

Your recommendations would be appreciated.

Dan Weber
Mercersburg, PA

Answer - Thank you for your question on upper entrances. Upper entrances are one of my personal favorites. I'll explain why.

Did you know that something like 80+ percent of all feral colonies have an entrance above the brood nest. Why?

Years ago I wrote an article in the ABJ about research I had done on Queen Excluders entitled "Queen Excluder or Honey Excluder." The research included information about upper entrances. What follows is a reprint of the last few paragraphs of this article as it addresses the colonies in this research that had upper entrances.

UPPER ENTRANCE COLONIES

As this experiment was proceeding I could tell that the upper entrance colonies were doing well. So, as a side note I would like to expand on the upper entrance theme based on observations made in the Queen Excluder experiment.

I am sure that many beekeepers have noticed that if an upper entrance auger hole is left open in summer or if there is a crack or gap between supers or perhaps a warped top, that a high percentage of bees prefer this entrance/exit. It was found that a large percentage of all colonies in the wild like to maintain an entrance above the brood chamber. One reason is because the very important brood chamber temperature can be maintained more efficiently than when exposed directly to drafts, breezes, from bottom openings. The slatted rack has often been proposed as a remedy to this problem in years past. But by just relocating the entrance to its more natural position, the expense and time needed to make, install and remove the slatted rack is eliminated. In fact, it is now my personal opinion that the only reason that we have hives with bottom entrances and a little front porch on the bottom board is because the early designers of bee equipment had front porches and doors on the first floor of their homes so "by George the bees will too!" If one has ever taken out a brood frame from the bottom brood chamber and closely looked at the brood pattern, it will be seen that this pattern is many times shifted towards the back of the hive and away from the entrance. In the Upper Entrance/Queen Excluder colonies this was not found at all. In these colonies it appeared that the sole restraint to the queen's egg laying was her ability to lay only so many eggs per day.

Another observation was that on very warm humid days the number of bees hanging outside at the entrance was much less noticeable than that of the bottom entrance colonies. Because of the upper entrance, the brood chamber was being ventilated by natural convection. The warm moist air was rising up and out of the entrance in a natural cycle as outside air entered.

As noted, there was a skunk problem that affected the results in the Queen Excluder experiment. The Upper Entrance colonies were left alone by the skunks, while the Bottom Entrance hives were fed on heavily at times.

I tried to mow this yard on a regular basis to control weed height as it affected the flight of the foragers. Weed height was not a problem with the Upper Entrance colonies as was the case with the Bottom Entrance colonies. Any beekeeper who has outyards and has had to mow or clip the weeds away from bottom hive entrances knows this is a problem area. Many trips to the outyards, and stings on the hands may be eliminated with the upper entrance.

I was much impressed with the advantages of the Upper Entrance colonies as observed in the Queen Excluder experiment. We as beekeepers are constantly barraged with information about how beneficial ventilation and moisture removal is in over-wintered colonies. The Upper Entrance is always suggested as a method to accomplish this in winter and in very warm humid conditions during the summer. There have been many, many articles and whole sections of books written on the upper entrance theme. The Rev. Langstroth's original book devoted a whole section to the benefits of the upper entrance and some of the most well known researchers in apiculture have also noted the benefits of the upper entrance. Perhaps we as bee-keepers should be more flexible, and look more closely at the Upper Entrance as a more efficient year-round option.

I, to this day, keep all of my colonies with year round upper entrances. I would suggest you do likewise. A copy of the complete article is available in the August 1985 ABJ.

BEES CLEANING CAPPINGS

Question - After extracting in the fall, is is safe to place the cappings 75 to 100 yards away from the beeyard so that the bees can clean them up? Will this cause robbing?

Answer - If the cappings are placed that far from the beeyard, I do not believe the bees would start robbing. The bees will clean up or rob the cappings and if hives are open in the beeyard while the robbing is going on, the robbing activity can extend to the hives. There is another thing you must be careful of and that is to be certain that all of the colonies were healthy and free of diseases. Know disease when you see it and make sure that none of the colonies from which the honey was taken, were diseased. If even one colony has AFB and the honey was extracted with the rest, the cappings would spread the disease to all the bees in the yard.

Question - When do I go down South to pick up package bees, I'm thinking of buying extra queen cells. I plan to divide a package into small units (1 lb. each) in order to have reserve nucs in each yard. What do you think?

A. Bzenko
Rochester, MI

Answer - A pound of bees is not very much, especially if you are going to try to get them going in spring in Michigan. I used to live in Michigan close to Grand Rapids and I think our last frost free date was June 7th.

Remember, this will work in warm weather, but it is pretty precarious in cold weather. There used to be a company that sold queen baby nuc "boxes" that used Ross rounds as the combs to keep the bees going. With warmth and food, it can be done, but requires constant attention.

PACKAGE BEES

Question - I have a couple of questions. When is the best time to start a colony of bees from a package? If I order Dadant and Sons Honey of a Hobby Kit No. 1 and the Companion Kit should I order bees at the same time?

Bill Polk
Stotts City, MO

Answer - *The best time to start a colony of honey bees, i.e. package bees, nucleus colony, split or swarm, is as early in the spring as your weather will allow. You want the weather to have settled somewhat and be free of continuous freezing temperatures. In your part of Missouri this would probably be the late part of March or early April. this early start gives your new colony all of spring, summer and early fall time to build up and store enough honey the first year to make it through the winter. You can order package bees from Dadants early the first week of March for shipment any time until the 1st of June.*

The Hobby Set #1 and Companion Kit take just a few hours to assemble, so it is easy to start beekeeping whenever you want.

THE SYRUP CAN AND THE QUEEN CAGE COMES OUT

UNCAPPED HONEY FROM POOR FLOWS

Question - What can be done to make bees cap honey filled supers? Also, the queen has started to lay eggs in the supers. What can be done? I have double-bodied hives with supers on top.

Answer - This sounds typical of a comparatively poor flow plus putting on too many supers. Many times, with a poor flow, the cells are not well-filled nor does the poor flow stimulate the bees' wax glands to produce an abundance of wax. Therefore, the honey cells are left open.

The fact that the queen is moving up to lay eggs in the supers is another indication of a poor flow. In a good flow the bees store the honey over the brood nest so that there are no cells for the queen to lay in. Now, as the honey in the lower part of the supers is used for brood rearing, the queen moves into the empty cell area and deposits eggs. I would guess that the bees are using more honey than they are bringing in as nectar.

The best thing to do would be to remove all empty combs from the supers. Then, concentrate the fullest combs of honey in one or more supers. The combs with small amounts of honey can be placed on top of the inner cover (after it has been placed on the top super of good honey combs). The hole in the inner cover should be open so the bees can go up into the supers with small amounts of honey in the combs. They will bring this honey down and help to fill the empty cells in the supers and brood area.

Question - I read an article in the January ABJ about using a vial of pheromone bee attractant in bait hives to attract swarms. I have had no luck locating this product. Where can it be obtained?

Fred Hardin
Lampasas, Texas

Answer - For the pheromone used in bait hives, try the company Phero-Tech., Mr. Doug McCutcheon, 7572 Progress Way, Rt. 5, Delta, B.C. Canada V4G IE9, Ph. 250-546-9870.

Question - I enjoy your column in the magazine. I use 6-5/8 super for brood box. I also use an upper entrance. Do I need to put a hole in each brood box or is one in the top 6-5/8 brood box enough? I use 3-6 5/8 supers for brood chambers and what size of hole should I use? --1/2, 3/4, 1"? Just one hole or 3 holes?

Don Barnard
Lincoln, NE

Answer - I have never liked drilling holes in perfectly good equipment. But, if you already have holes, a couple more won't hurt I guess.

I make use of a 3-sided spacer to establish an upper entrance without drilling any holes. This fits between the upper hive body and the first super. Seems to work rather well. Your hive covers can also be propped up with a shim or stick to provide an upper entrance and escape for water vapor generated by the clustered honey bees.

Question - I have three hives and they are setting quite close to my garden (20 ft.). Sometimes when working in the garden the bees want to sting me. My question is, will I be able to move these hives back 10 or 15 ft. of even more in the dead of winter without disturbing them?

Michael Golden
La Plata, MO

Answer - Yes, you can move the bees during the winter. The main consideration is if the weather is below freezing and the bees have formed a tight cluster, you don't want to severely bounce or jostle the bees. The possibility exists that a strong jolt could dislodge part or all of the cluster. This part may then not be able to reform and freeze to death. If the queen is in this part, then your colony is in real jeopardy.

If you have some help to lift the hives, the short distance you are going to go shouldn't be a problem.

PREDICTING THE HONEYFLOW

Question - When can I expect the honeyflow in my area (Illinois) and when should the honey supers be taken off the hives?

Answer - I would assume that the flow in your area would not be much different than it is here at Hamilton. Your floral sources are probably very similar to those of our area as well. White Dutch clover is an early yielding nectar source and contributes to good honeyflows from late May on into June. White Dutch overlaps the flow from yellow sweet clover in early June. Likewise yellow sweet overlaps white sweet clover a couple of week later. White sweet clover continues into July.

It is advisable to remove your supers of white honey from the clovers by early August to prevent it getting mixed with the darker honey from fall honey plants such as Spanish needle and smartweed. These fall plants give the bees some honeyflow but normally not in quantities worth taking off. in addition, these later fall honeys are darker in color and stronger in flavor and are best used for winter stores.

ENTICING BEES INTO SUPERS WITH FOUNDATION

Question - I am having problems persuading my bees to work in an empty super of foundation above a filled honey super. What can I do to entice the bees into this super? The first hive body is being used for brood rearing and the second hive body is completely filled with honey. After that, I have a queen excluder on and then a full super of honey and the empty super with foundation.

Answer - Persuading bees to work foundation is often a problem, especially if a heavy honeyflow is not currently on. There are a couple things you might try, however.

First, I would suggest removing the queen excluder since the second hive body is full of honey and it will act as a barrier to prevent the queen from moving up into your honey supers. After removing the queen excluder, reverse the full super with the empty super so that the empty super of foundation sets directly on the second hive body filled with honey. Then on top of that put the full super of honey or go ahead and remove it for extraction. Without the queen excluder barrier the bees will be more apt to begin drawing the foundation and they may even move some of the honey from the second hive body into the super thereby allowing more brood rearing room.

Still another idea used by some beekeepers is to provide an upper entrance in addition to the normal bottom entrance. This will allow easier bee access to the top empty supers.

Question - On page 634 of *The Hive and the Honey Bee* a horizontal two-queen system that was developed at the University of Minnesota is mentioned, but not explained.

Will you please give an explanation of the horizontal two-queen system? Thank You.

Paul Judd
Cottonwood, AZ

Answer - *As you probably know, two-queen systems either stocked with one queen inside a hive body on top of another queen inside a hive body or horizontal with queens side by side in hive bodies or similar confines have been around for a very long time. The theory with a two-queen system is that, bottom line, they collect more nectar than two separate colonies. In practice, setting up a two-queen colony is relatively easy to do on a basic level. Like everything else in beekeeping, though, there are ways of micromanaging this type of colony for optimum results.*

Dr. Marla Spivak is now in charge of beekeeping interests of the University of MN. This is what she shared with me.

"We published a manual of how to do Dr. Furgala's system of management. It sells for $15. We also just put out a video to go with the manual, the video alone is $40. With the manual, the package is $45.

"We also teach a 2-day short course on the method every March. The course is $45, and includes the manual (and lunch). If you take the course, the video is $30.

"So, if you're interested in any of this, send me your full address, and I'll send you an order and/or registration brochure." Dr. Marla Spivak, Dept. of Entomology, University of MN, 219 Hodson Hall, St. Paul, MN 54108.

**CHECKING THE QUEEN TO SEE
IF SHE HAS SURVIVED THE TRIP**

Question - In spring when I start to feed my bees and put pollen patties (Pollen substitute and sugar syrup) in the colonies for them, the patties always dry out and get hard. I would like to know what is the best way to prevent this from happening so the bees can use it?

Earl T. Williams, Jr.
West Monroe, LA

Answer - In the past I had the same problem as you with my pollen substitute patties drying out.

I found two ways to fix the problem. I took a hint from my wife and added about a tablespoon of liquid vegetable oil per patty. This kept the patty, when well mixed, firm and less prone to drying out. The reason it was drying out was that my sugar syrup had too much water in it and not enough sugar. When the water evaporated it left a dried out patty.

Another way to solve the problem is to make a thicker sugar syrup or use straight corn syrup. I use straight corn syrup now and have much better accepted and used pollen substitute patties.

Question - We never had any previous experience with bees. We obtained a colony of bees last summer. For the winter, my husband covered the entire bee hive with a heavy plastic sheet for winter protection. Now we found out all the bees are dead although they still had some honey left for food. My husband contends they died of starvation. I feel they died of suffocation. How important is ventilation? I am most unhappy over what happened. Do you have any answers, please?

Arthur Madden
Valley City, N.D.

Answer - The death of your colony was probably a combination of suffocation and stress from high humidity and CO$_2$ levels.

Bees eat honey as a food over winter. If the honey is at 17% moisture and your bees consume 100 lbs. of honey, they liberate 17 lbs. of water. If you wrap them in plastic, then this water cannot be removed into the outside atmosphere, the oxygen they need as living creatures is limited and the waste product carbon dioxide is at higher levels than that which is optimum. You wouldn't put your dog in his doghouse, stuff the doghouse with food and water, then wrap it in plastic nor would you put a plastic bag over your head for any reason.

The hard lessons are always the most memorable. Don't be discouraged. Look forward to next spring.

Question - I have just read about "Fuller Candy" for feeding bees. It is supposed to be so much better than feeding sugar syrup. No chance of encouraging robbing and is better to handle, etc.

Can you tell me about this and how to use it and where to go?

Kim Best
Provo, UT

Answer - The Fuller Candy recipe makes a soft fondant, similar to that in chocolate creams. It may be made to the degree of hardness or softness one prefers. The higher the temperature the candy is boiled, the harder it will become.

INGREDIENTS:

12 pounds granulated sugar
1 1/2 pounds liquid glucose
1 1/4 quarts water (5 cups)
1/4 teaspoon cream of tarter,
added when temperature reaches about 230º F. (110º C.)
Boil to 238º F. (114.4º C.)

As soon as the sugar begins to dissolve, prior to boiling, the spoon or paddle used in stirring should be removed from the kettle. Do not stir candy while cooking, as this will cause a coarse grain. Boil to 238º F. (51.6º C.), stirring vigorously until the mass appears in color and consistency to be like boiled starch. At once pour into molds or feeders and cool.

Question - This year I am planning to expand my colonies by purchasing package bees. One problem has arisen in which I have researched, but have not been completely satisfied. I want the package bees to produce a surplus of honey. However, (the problem) I can't pick up the bees until the 4th week of April. Considering normal attrition of the adult bees for the first 3 weeks and not having enough brood frames for all the package hives, I have been studying an idea. Your comments/suggestions are most welcome.

(1) Purchase both 3 lb. and 2 lb. package bees, half and half. (All hive boxes have drawn comb.)

(2) After the queen begins laying and about 8 to 10 days later, transfer 2 capped brood frames from the 2 lb. hive to the 3 lb. hive. Replace with empty drawn.

(3) Prior to all the brood hatching, unite the 2 lb. hive on top of the 3 lb. hive. Remove the queen to replace other queens doing poorly in other hives. Each hive box contains an in-hive feeder to further stimulate growth.

It would seem that the queen in the 2 lb. hive will continue to lay until the two hives were united, insuring more brood frames for the 3 lb. hive, perhaps doubling the population of bees normally expected in the first 45 days. Considering all things equal, would this approach insure the growth I am looking for, to expect a modest surplus of honey?

Gerald Fields
Holts Summit, MO

Answer - You have identified the main problem of obtaining an early honey crop from package bees – lack of bees until the 21st day. In fact some beekeepers only order 3 lb. packages, buy "booster packages" or purchase nucs to speed up colony buildup before the main honey flows.

Having a source of additional developing brood will certainly boost your hive capabilities. I can't promise you a honey crop, but you will certainly have more bees to forage for one.

Since you will have access to two queens, you may want to experiment with two-queen colonies. You may want to research two-queen colonies and increased honey production in the literature. There is an excellent discussion of this topic in The Hive and the Honey Bee textbook available from Dadants.

PARTLY FILLED FRAMES IN UPPER STORIES

Question - At the end of each honey season each of our hives contains a great many partly filled honey frames. Most of these frames will have from 5 to 50 cells filled with capped honey. We will have so many partly filled frames in the higher supers that in the fall it becomes difficult to cut the colonies down to two or three stories. How can we force the bees to remove this honey to the lower stories?

Answer - You could place the frames containing the cells of honey together in some supers, put one of these supers on top of a hive with an inner cover between the hive containing the bees and the super containing the honey, the hole in the inner cover being left open. The cappings on the honey should be broken with a cappings scratcher and the bees will carry the honey down. All supers should be removed from the colonies first, and the inner cover and super with the honey placed directly on top of the hive.

If you are sure all of your bees are free from disease, the cappings on the honey can be broken and the frames set outside where the bees can get to them. They will rob the honey out, but that is a good way to spread disease so be certain your bees are free of disease first.

It would seem to us that you put supers on your bees too rapidly. According to your letter, you have several supers on each hive with scattered cells of honey in them. The second super should not be put on the bees until the first super is nearly full of honey.

Question - In the course of removing honey supers, I have lost a queen to one of my better hives. At this time of the year what can I do?

Robert Sparks
West Frankfort, IL

Answer - I'm sorry to hear that you lost a queen at this time of year especially. Because cold weather has closed in on us and the lack of available queens from the southern breeders, this colony probably can't be salvaged as an individual hive.

My suggestion is to use the extra bees and honey to give a boost to some of the hives that you may have that are a little weak. You might as well use these bees and honey to bring other colonies through winter stronger for spring.

Question - I am an amateur beekeeper and would greatly appreciate any help, advice or suggestions with regard to the problem I recently encountered as follows.

I fed artificial pollen to one of my hives using this formula: I made a mixture of 2 parts sugar, 1 part water and 1 part artificial pollen which I cooled. I put the resulting putty on top of the frames.

After a few days, I noticed many dead bees in and around the mixture. I removed this blend and the bees stopped dying.

A few weeks later, I duplicated this feeding method with a different hive to verify my findings. Unfortunately, the results were the same and a quantity of dead bees were grouped around the mixture.

Thank you for any assistance in this matter with regard to a possible solution to the difficulty.

Lauro Guerrucci
Decatur, GA

Answer - Thank you for your question. I'm sorry to hear about the results you had in trying to feed a pollen substitute. A simple answer to your question is that your mix-

ture was simply too sticky or wet and your bees got stuck in it and died. I have often seen this happen with bee feeds.

I think we can eliminate the sugar and water as toxic ingredients, assuming they were fit for human consumption, we will make the leap of faith that they were also fit for honey bees. The great variable is the "artificial pollen" that you state you used. You never made mention of what it was. Most pollen substitutes are made from expeller processed soybean flour and sometimes brewers yeast and whey fermentation. There are products which bees will feed on that are all right for you and I or other mammals, but which are toxic to honey bees. So, if you would not mind, please let me know what the artificial pollen part was and where you purchased it?

In your letter you mentioned that you let your mixture cool. There is also the possibility that if the "heat" was so high that the protein or other compounds in your mixture may have been changed or altered in a way deleterious to honey bees.

In any event, if you could give me a complete ingredient list and where it was purchased from, where it was stored, what it could have been exposed to, etc., perhaps we can shed more light on the problem.

Question - I have a question: I have 7 hives - the bottom box on each hive stand is a 9 1/8 deep for brood and the next box on top of that is a 5 5/8 shallow for brood. Well, things started looking crowded so I added a 9 1/8 deep on top of that. The queen immediately started laying up in the top 9 1/8 deep. I feel like I made a mistake by using the 5 5/8 shallow for brood instead of a 9 1/8 deep. I think the queen needs two deeps for brood. My question is: Is there any way of correcting this problem with these hives during this season? I've been told the ideal set up for this area is 2 deeps for brood chambers, then a queen excluder, and then more deeps for honey. Thank you for your help.

Mark Gosswiller
Boise, Idaho

Answer - *If one were talking strictly about space for brood rearing only, one 9 5/8 with ten frames with perfectly drawn comb of only worker sized cells would be necessary for the brood chamber. Even if you factor in a band of pollen and nectar above the brood for feed, then a 9 5/8 and a 5 5/8 inch box would be okay. The problem arises, especially at this time of year with package bees, that their population expands faster than the beekeeper has anticipated. As a result, this growing population brings in nectar very fast and stores it in any available cells. These cells may turn out to be ones the queen could have also laid eggs in. Now she can't. You then added more "space"; in this case another deep and the bees respond by using it. They're not dumb.*

For your area and climate two brood chambers (one for brood rearing and one for honeystores) and a queen excluder then supers would be the best compromise at the beginning of the season. Right now you can locate the queen and move her below or switch brood chambers if she is above, moving her below with the whole brood chamber. Separate with a queen excluder the brood chambers from your 5 5/8. If there is brood in the 5 5/8, that is not a problem. The brood will develop and emerge, leaving you with a super that is likely empty except for some pollen still stored in the cells. The bees will then refill these empty cells with honey.

SUPERING

Question - I would like to know why bees may not fill out the second super, but will fill the first?

Answer - There could be any number of reasons why the bees would fail to fill out the second super, but most likely what has happened is that the honeyflow has ceased or essentially ceased before they had a chance to get at it. This, of course, would be the most common answer. Other possibilities would be that your field force may have been destroyed by a pesticide application and, therefore, you had no bees to bring in the remaining nectar.

SUPERING

Question - I have 6 hives of bees and I gave them two boxes for winter. Can I add 2 more boxes at one time in May? A party told me that is what he does and that is all he adds. Where should I put the inner cover? Leave it on the 2 boxes or put on top of the 4 boxes when I add the two? Also would like to put on a few small boxes for comb honey, 1-lb. boxes, as I have some supers for comb honey filled with 1-lb. boxes. Where should I put them?

Answer - In May, you can add two or more boxes or supers at one time and it may be plenty of space, but we usually find that bees should be checked and supers added as needed. They seldom get away with only two supers. The inner cover goes on top of the stack under the cover.

   Bees, as a rule, do not efficiently produce both comb and extracted honey at the same time. They produce one or the other in a single colony. Bees should be confined to one hive body during the honeyflow if they are to produce the best comb honey.

Question - I have a question for the classroom. I was able to collect 8 swarms this last summer and had much success in it. But one of the swarms must have come from some place other than another hive, because it took two hives to bring it home. My problem is this: The weather was warm (in the low 90s) and the vehicle I used didn't have any protection from direct sunlight. By the time I got the bees home and set up in the back yard (a trip of about 20 minutes) this massive swarm had died. There were dead bees 4 inches thick on the bottom of the hive. There was wetness (something that I took as being everything these little helpless bees had inside them) everywhere. To say the least I was extremely sad and devastated. I had accidentally killed one of nature's most beautiful sights. Now my question. How do I transport a large swarm in inescapable heat?

Mark Gosswiller
Boise, ID

Answer - *If the heat is inescapable don't collect the swarm. Let them go to wherever swarms go. I'd rather do that than cook one.*

Let's look at alternatives. There are going to be variables based on each individual circumstance. When it's hot and humid you may want to design a large screened box to dump the swarm in for the short trip home. This can be shaded without closing them off and thus allow maximum ventilation. What you don't want to do is collect a swarm in this kind of weather in a sealed container like a hive. Substitute a screen cover instead of a regular one, turn your air conditioner on high or your outside vent or roll down all your windows or do all these things – ventilation, ventilation, ventilation.

I've never tried this but it might work. Stop at your local convenience store and buy a couple bags of ice. Put a bag right in with the bees if there is room.

Commercial package bee and hive transporters use netting on their loads and they drive straight through to their destination by using two drivers. They are equipped with water sprayers to cool bees down in emergencies. Some have used refrigerated trucks to transport bees.

Question - In extracting my honey I found twenty frames of honey that were full but not capped. I froze it. Can I take it out next spring and put it back on the bees so they can use it?

Thomas Mulinix
Kingston, GA

Answer - *The honey in the frames that you have frozen will be fine for spring feeding. You did the right thing. Your bees will make good use of this extra feed in the coming spring.*

Question - I'm starting out as a beginner. I want to work long term as well as short term. What do I want to do next year? Expand. It would help if I have drawn frames for my plans. Can a learning beekeeper continue to feed the bees to keep them expanding, filling and drawing foundation, then extract the frames and feed the extracted sugar syrup back to the bees and store the drawn frames for the next season?

William B. Miller
Cleburne, TX 76031

Answer - *Yes, you can force the bees to continuously draw out foundation just as you describe. Remember also that with lots of sugar syrup and natural pollen coming in that the queen will be laying wall to wall if given room. This means, of course, a <u>large</u> population of honey bees that may want to swarm. You may want to think ahead and plan on splitting your colonies to create even more if you don't want a honey crop, at this stage.*

Question - We would like to know if wood smoke is harmful to our dormant honey bees?

Our farmer neighbor on top of the hill to the northwest of us has one of those outdoor wood furnaces. Morning and evening he stokes it up and if the wind is out of the NW (which it usually is in the fall and winter) the smoke rolls over his house and down the hill like fog. It envelopes our house, property and bee yards with smoke; at time it stings your nose and eyes.

We have noticed on warm days more bees than normal seem to be coming out of the hives. Of course they get chilled and don't make it back to the hive. I wonder if ... the smoke could be keeping the bees upset?

Mrs. J. Damron
Stephenson, MI

Answer - Any living creature that is forced to breathe smoke of any kind is negatively affected by it. In your situation I can positively tell you that the smoke is not good for you or your bees, but concerning the bees I can't tell you how bad it is.

In regards to honey bees, smoke both masks the alarm pheromones produced when a hive is disrupted and alerts the bees, causing them to gorge themselves with nectar or honey in preparation to abandoning the hive. Repeated long-term exposure must be disruptive to the colony. How long this disruption lasts or what the long-term results of this disruption are, I cannot answer.

Your report of "more bees than normal" leaving their hives could be a combination of smoke, mites, other disease, weather, sunshine or just coincidence of your closer observation. Hopefully, your populations are good enough to handle these extra deaths.

Question - An instructor at the University says bees can't digest ground corn so they don't get any good out of it. He says they haul it in and haul it back out. Is this true?

Pat Chandler
Anselmo, NE

Answer - The instructor at the University is right that bees cannot digest large chunks of ground corn. It's not really that they couldn't digest it if they could get it in their digestive system. The pieces are too big. Grind this corn into a dust with particles sized comparable or smaller than pollen grains and they will eat and digest it. Back in 1967 a researcher by the name of M. Haydak in Minnesota tried feeding just about any protein-containing food you or I eat to honey bees. You get the particle size right and they will consume just about anything. Corn flour he found was not a very good protein source for honey bees because its amino acid structure is not complete for honey bee nutrition. Haydak found that a mixture of soybean flour, dried brewer's yeast and dried skim milk was the most effective pollen substitute.

Question - Steve Taber's article on foraging areas in the September issue reminded me of a question that I planned to ask you already a long time ago. What is the right size of a single apiary site? How many hives (strong Buckfast colonies, with two deep supers as brood chamber) would you put into one site? And–concerning the foraging area–what is the minimum distance between two sites?

In my area–woodland and dairyland–there is a low to medium nectar and pollen flow throughout the season with one or two quite unpredictable peaks. Average yields per hive vary between 20 and 150 lbs. in different years.

Thomas Kober
Germany

Answer - Thank you for your question on foraging areas. I wish there was a nice neat black and white answer for you. Unfortunately, there isn't, but let's look at the variable, which affect the situation.

The size of any apiary, anywhere in the world, is dependent on several factors: (1) Nectar and nectar sugar concentration available within optimum honey bee foraging distance - 1 mile or less. (2) How many nectar foragers are competing in the area? This includes honey bees, bumblebees, solitary bees, beetles, butterflies, flies, etc., etc. (3) Weather plays an important role. Is there enough moisture, heat, daylight, no wind, little humidity, etc., etc.

In a perfect world, you could put 500 colonies in a 500 acre sweet clover field in a hot, dry summer and have a excellent honey crop. But it's not a perfect world and your colonies are in a mixed woodland, dairyland area. As a pure guess on my part, I would think 10 to 20 colonies no closer than 2 miles apart, depending on your assessment of the above variable, would be the maximum carrying capacity of your area.

This was a excellent question and if we were working with a mathematical question instead of a fluid environmental question, a more substantial answer could be derived. But we have our continually interesting and variable honey bees with which to work with. That's probably why we are beekeepers–the challenge.

Question - I provide overwintering honey for my bees by feeding them sugar syrup in the fall, having extracted as much of their honey as I can. I deliberately overdo the feeding quantity, preferring to have to much rather than too little. As a result, I wind up with a lot of unused overwintering stores from syrup in the spring. The problem is that a lot of it, probably most of it, has crystallized.

My question is whether or not I dare to utilize this crystallized syrup for their overwintering stores next fall? Can I feed dry granulated sugar? Also, I wonder if there is another source other than granulated sugar to use for syrup? High fructose corn syrup, perhaps? It would have to be available in smallish quantities to be practical for a hobbyist's few hives.

Dan Hendricks
Mercer Island, WA

Answer - *I'm going to assume a couple things. First, that your bees are kept in close proximity to your Mercer Island address and second, that your temperatures in the dead of winter are nowhere near as low as they get here in West Central Illinois. If I can assume those things, I'd say feed the granulated combs of sugar syrup and the dry granulated sugar. There should be plenty of moisture for the bees to use to reliquefy the sugar themselves in their general environment and without long periods of sub-zero weather, they should be fine. You may have to uncap the frames of crystallized syrup to get the bees started on it. Another trick is to soak the crystallized frames of syrup in water for a day to loosen the crystallized sugar.*

High fructose corn syrup (the bee feed most used by commercial beekeepers) granulates also, so it's not really a solution for your crystallization problem in the spring. The only other option you really have is to not take all the honey. However, even leftover honey sometimes granulates by spring. But, you wouldn't have to worry about possible sugar syrup adulteration of honey in the surplus supers.

Question - How do I get my bees from old equipment into new equipment?

Kenneth Dreese
Larned, KS

Answer - *There is really no easy way to transfer bees into new equipment. The way that is potentially the easiest, is to add an additional hive body and frames and wait or hope that the colony will begin to draw new comb and the queen will move up into it. Depending on a variety of possibilities, this sometimes works quickly, sometimes slowly and sometimes not at all. The surefire way is to tear into the old equipment, find the queen and move her into the new hive body, then put a queen excluder between the old hive body and the new. This keeps the queen in the new equipment and out of the old equipment while allowing whatever brood is in the old equipment to emerge. After at least three weeks, you can remove the old hive body and shake the remaining bees out into their new home.*

Question - Our hive consists of two hive bodies. The upper body appears to be well filled with honey. We took off about 20 pints of honey from the two supers. The lower body was used for brood, but may have some honey in it by now.

Does that seem to be sufficient honey for the winter? Would it be a good idea to check the hive in mid-spring and if short of honey, feed sugar water?

One pamphlet shows a hive consisting of one hive body and one super with a one inch hole bored in the super under the hand hole. Should I bore such a hole in our upper hive body? If so, should it be screened to prevent mouse entry? Also, if so, should the hole be plugged for summer? Our hive faces south on a gentle slope, but is exposed to strong north winds. Would it be a good idea to place a windbreak on the north side?

Lastly, we didn't get all the honey out of our two supers, although we warmed them in the furnace room before extracting. Is there some way to get all of the honey out of the supers? Will it be proper to place the supers back on the hive in the spring with the leftover honey in them?

John Munro
Fayetteville, N.Y.

Answer - For most of the northern part of our country, yours included, it has been suggested that 60 to 90 lbs. of honey are necessary to both provide not only enough carbohydrates for the bees during the winter, but also to provide enough for early spring to allow brood rearing to take place. Brood rearing starts in late winter and early spring is the biggest user of both honey and pollen. If you think towards the end of winter or early spring that stores are being exhausted, then feeding a thick sugar solution with as little water as necessary would be wise. Remember, the more water put in, the more that will be released in the hive, perhaps resulting in excessive condensation.

I don't think I would drill a hole in the upper hive body. I've always thought this a little extreme to drill holes in perfectly good equipment. The reason always given for drilling a hole in upper hive bodies is to allow the water released by the cluster's metabolism to vent out, like a chimney, and to give them an upper entrance/exit when snow or dead bees obstruct the lower entrance. Instead of breaking out your Black & Decker, simply prop a corner of your hive top/inner cover up about 1/4" with a small stick placed sideways between the upper body and the top.

This question brings up some additional thoughts. Why, as beekeepers, are we always trying to ventilate a colony in winter when the bees have spent several weeks in fall collecting propolis to seal it up? Are we, in effect, helping the bees because they are to genetically deficient that the knowledge and action to open up and ventilate their home has somehow been lost or missing. Or, are we stressing our bees more by interfering with their action to seal up and make their home as draft-proof as possible?

Any windbreak you can provide would be a good idea. Wind chill can be severe for the bees as well as you!

I would do as you suggest and put the supers back on in spring to give an additional food boost at this time of the year.

LURING BEES INTO HONEY SUPERS

Question - I have a hive of bees that doesn't want to fill the super above the brood nest. What can I do to lure them up into the super?

H. Chin
Beardstown, IL

Answer - Try using a bait comb to lure them up into the super over the brood nest. This means taking a comb or two from a strong colony where the bees are building comb and storing honey. Place these combs near the center of the super you want the bees to occupy. The frames of foundation from the non-working colony can be put back in the colony from which the combs were taken. The bees of the non-working colony will then be more likely to be attracted up into the super where the two combs with honey in them have been placed. Generally, they should continue to work on the adjoining frames containing foundation and should soon be working much better. Also, if you use a queen excluder, remove it until the first super is being worked actively by bees.

Question - I would like information from anyone concerning the problem I have had recently with 4 hives. I have 8 hives total, all Italian except 1 Starline. All hives had 4 supers each. I had checked all hives about 2 - 3 weeks before I noticed a problem. At that time all were full and ready for harvest. I noticed a lot of dead bees and chewed up beeswax in front of the 4 hives in the north row. I found most of the bees dead, all honey gone and frames really riddled. The other 4 hives had not been invaded at that time. In cleaning up afterwards, I found a lot of small black bees, all dead. After talking with some of our local beekeepers, no one had heard of such an invasion. Some suggested it may have been wild bees from a bee tree in the area. I know of one tree about mile from my location. The owner of the tree called me about 2 months ago after a rain storm had caused bees and comb to fall into his driveway. There seemed to be a lot of comb building with bees in some of the larger upper limbs of the tree. I told the owner he had three choices, to cut the tree down, call a pest control company or keep the tree. Could these bees have been my problem.

Virgil Goff
Ft. Smith, AR

Answer - *It's hard diagnosing from afar, but from your description I'm going to guess your colonies were robbed out.*

Let's guess at a scenario of what happened. Your bees appeared well and healthy with a good population for most of the summer. They stored some honey and everything was fine. Then, quickly and dramatically from either mites, disease or queenlessness, the population of these colonies dwindled. They dwindled to such an extent that other colonies of your own bees, someone else's bees in the area or a colony in the wild saw an opportunity – an opportunity for free food. They overwhelmed the colonies, killed off the rest of the colony and took the stored honey back to their own colony. Once the robbing starts, all the bees in an area will participate. It's every bee for herself.

If this is the reason, it gives added impetus for all of us beekeepers to really check our colonies more often than in the past. With mites especially, population can drop quickly and things like this happen.

Hope you had better success with your other colonies!

Question - I have a hive consisting of two hive bodies that refuses to draw out the comb or store honey in my supers. The hive is aggressive and I plan to requeen next spring. My question is why won't they work in my supers, it's almost as if they know that I will take it from them.

Also, instead of paint, could I use a waterproofing wood sealer such as Thompsons on my hives?

One last question, I heard the other day that there is another mite coming that is worse than the two that we have now. Is there any truth in this story?

Colby McMullen
Williamsville, VA

Answer - Sorry to hear that your one hive is not doing anything. Several possibilities exist, but the bottom line is that you have to have lots of bees that go out and collect lots of nectar which needs comb to be stored in and need additional storage room. If the bees have swarmed several times without you noticing, mites, disease, etc., which keeps populations down, then if there is little or no additional nectar above their needs, they have no need to draw foundation to store honey. I don't think they realize that you will take the honey from them. I think that there just aren't enough bees with enough nectar to progress.

I think Thompsons on the outside of your hives would work great. It's a little expensive, but if your woodenware lasts longer, it may work out financially for you. This should not hurt the bees or honey. Let it dry thoroughly. Some beekeepers use Thompsons or another wood preservative first and then add a coat of paint.

I hope it is not coming, but there is another parasitic mite, as you say, named Tropilaelaps clareae. The distribution of this mite is still relatively restricted to the area around southeast Asia. The mite is a little smaller than the Varroa mite, so is a little harder to detect with just the naked eye. It will infest colonies already infested with tracheal and varroa mites, but it is equally sensitive to fluvalinate (Apistan) as a control. Natural distribution and expansion of this mite would take decades to reach North America if at all. But some irresponsible beekeeper could collect bees or queens from this part of the world that are infested with this mite and smuggle them into this country in a matter of hours if he/she so desired. Cross your fingers!

Question - Have there been any experiments on wintering bees in combless packages either indoors or outdoors? If so, what were the details and results? Thank you for any information you can give me.

Rufus Payne
Apalachia, VA

Answer - As far as I know there have not been many long-term studies to see how long honey bees will live in combless conditions. Honey bees are so closely tied to comb in their whole life cycle, it is difficult to separate one from another. They use comb as a life support warehouse if you will. It is their home—where everything from protection to raising young from egg to emergence, honey and pollen storage occurs. It is where the queen begins existence and drones are propagated. The first things bees do in the wild is start building comb, even if it is out in the open and unprotected by the hollow cavity of a tree. I've even seen package bees begin comb building in their container as it is shipped through the mail. Without comb, honey bees do not exist long-term and that is why their genetic tie to comb building and repair is so strong.

I'm not saying it couldn't be done, but smarter minds than mine have not found a reason to study it as yet.

COLONY MANAGEMENT

Question - Is it necessary to have more than two supers per colony?

Answer - Yes, many times in a very fast nectar flow a good colony of bees will have nectar scattered through as many as four supers and not have any of it cured and capped, and ready to take off. Therefore, if you do not have the extra supers, you will lose part of your crop.

MAKING NUCS

Question - I am planning to make nucs for increase this spring. I plan to make the splits around the 10th of April and I would like to know how many weeks ahead I should begin stimulating the parent colonies in order to build them up for splitting?

James Neagle
Richmond, VA

Answer - If you are planning on making up nucs as splits the 10th of April, you will want to start stimulative feeding of sugar syrup and pollen substitute six to eight weeks in advance of this date. Assuming that your weather will be mild enough so that the bees will consume this feed consistently over this period of time, your population of bees should be large enough to make strong splits.

Question - I have three questions concerning early spring stimulation:

1. I have a considerable amount of old honey from my bees which is disease free as far as I know. It is a dark thick tulip-popular type. Is 50% water and 50% honey the proper mixture? If not, what is the proper mixture?

2. I have always mixed 5 lbs. of sugar to one gallon of water; now I read 1 lb. of sugar to 1 pint of water. The mixture I use seems to be by volume while the other seems to be by weight. Is either one O.K?

3. Some beekeepers simply scratch combs instead of feeding. Assuming the hive has ample stores, is scratching combs as good as feeding sugar syrup for early spring stimulation?

Jim Neagle
Richmond, VA

Answer - (1) I would say that a 50/50 mixture of honey and water would be fine. But why not just feed it straight and by-pass the mixing? The feed would be concentrated and not need moisture removal by the bees.

(2) In cold or cool water 5 lbs. of sugar will dissolve fairly well in 1 gallon of water. Heat that water up to 110º-120º F. and you can fit 3 more pounds in to equal 1 lb. per pint.

(3) For early spring stimulation either one will work well. When you feed anything from a feeder in the hive, each individual bee has to find the feeder by themselves as the bee's dance language doesn't have any "words" for anything closer than about 3 meters to the hive. It takes a while sometimes for all the bees to find this resource. Scratching combs is faster and doesn't require labor, feeders or sugar.

WINTER BEE FLIGHT

Question - Here in Union, Missouri I have noticed my bees flying on very cold but sunny days. Is this normal?

Kim Sieve
Union, MO

Answer - Yes, in small numbers this is normal and nothing to worry about. But it is interesting because it brings into question microclimates. Have you ever noticed in winter on the coldest but sunny days heat waves radiating off the black pavement of the road you're driving on? Well, even though it's very cold "generally" many times on sunny days the temperature is very warm even tropical several inches above the surface of a road or landing board of a hive in your backyard. As a hive collects solar energy on a day like this and warms up some, bees venture out and if they stay close to the ground or structure that has been warmed up, they are fine. But if they venture out of this zone they chill and can't make it back to the hive.

WINTERING AND GRANULATED HONEY

Question - I have a few questions about keeping honey bees in the desert region near Las Vegas, NV.

(1) Wintering, how much is too much in the desert? I've counted and last winter (a cold record breaker for Las Vegas) my bees were flying all but 6 days. The previous winter I wound up with about 8 frames of honey in one hive in February. I had only left 6 frames in the previous October, and note very little use of honey I leave for the bees to winter with. How then is this going to affect feeding for spring buildup and what time should I expect the buildup in the dessert?

(2) What's the best time to harvest the honey and what do you do with desert Tamarac honey that is sugared in the comb?

John Allen
Las Vegas, NV

Answer - Keeping honey bees in winter in Las Vegas has some definite advantages. The bees can fly most of the year which helps in the overall health of the colony. Here in Illinois the bees cannot fly for weeks, sometimes months at a time and therefore, nosema is a problem along with chalkbrood. These are not your problems in the desert. Bees are able to forage on flowers most of the year. Therefore, having many pounds of honey stored for the winter is not necessary. Spring buildup starts in full force in late February or early March. This coincides with the blooming of many ornamental plants and catsclaw acacia, ironweed, etc., which are indigenous.

You can harvest your crop of honey whenever it has been capped and/or you are running out of storage room. Any honey that has crystallized in the comb should be left and used as feed for the bees later. They can remove it and use it to some degree. The beekeeper can't. Be sure to check you supers often when you know your bees are bringing in nectar that will crystallize as honey and extract it when the comb is 80-90% capped.

COMB REPLACEMENT

Question - My present hive has three frames which are almost solidly connected with comb (they weren't put in straight). Should I break up the comb and put in new foundation or leave it as is?

Cary Conway
Visa, IL

Answer - Move the three crooked combs to the outside positions in the hive. When there is a nectar flow in June and the bees are drawing comb readily, replace them with frames containing straight full sheets of foundation. If the crooked combs contain honey, place them in a body on top of the colony and above the inner cover with the hole open. The bees will soon take the honey down and dry the crooked combs. They then can be removed and the combs melted down into wax.

Question - Is it important to use entrance reducers during the winter and when should they be put on?

Rod Little
La Harpe, IL

Answer - Entrance reducers reduce the amount of cold air that can enter the hive and will keep the mice out if covered with 1/2 inch mesh hardware cloth. Mice will start to move with the first freeze of fall, coming in from the meadows. Many will make their way into a hive if nothing restricts them. They will go back into the corner of the hive, make a nest of grass, and do much damage to honey and the combs. An entrance cleat can be made by sawing a 1-inch board long enough to fit into the entrance, then make two cuts about 4 inches apart near the center of the board and 3/8 inches deep. Chisel out the part, making an entrance 3/8 x 4 inches. This will keep all mice out, except the shrew, which is somewhat smaller than the field mouse. Covering the entrance portion of the entrance cleat with hardware cloth will ensure both are kept out of the hive.

Question - Towards the end of every season I have two problems that should have a common answer. The first is honey supers that are only partly filled that I want the bees to take down into supers that are almost fulled and capped. The second is "wet" supers that I want the bees to clean up while placed on top of hives.

I know that I can set the supers aside and let the bees rob them out, but I would prefer not to do this as it may encourage other undesirable robbing and it gives wax moths a chance to lay eggs. I have tried placing an inner cover between the hive and the supers to be cleaned with the square hole for a bee escape left open. Also, instead of the inner cover I have placed on a box feeder. In both cases the bees assumed the to-be-cleaned supers were theirs and carried on working in them.

Is what I want to do possible? If so, what do I have to do?

John H. Nelson
Almonte, Ont. Canada

Answer - I have the same problem as you with partially filled supers and "wet" ones and what to do with them. If we could just communicate rationally with the bees and tell them what we would like them to do and why, it would be good to do it. It certainly would make the beekeeper's job easier.

I'll tell you what to do and maybe give you some ideas. Wet supers: I take them about 100 yards form my apiary and let the bees clean them up. This is far enough away to discourage any robbing behavior in the apiary. I've tried putting wet supers back on the hive to be cleaned and it just doesn't work for me. I don't worry about wax moths laying eggs in the supers because they do it at night in a hive full of bees anyway and as long as I take precautions in how they are stored, no harm is done.

Partially filled supers can be included with the wet supers and cleaned out or combine partially filled frames together in supers and use them as early feed in the spring for your colonies. This stimulus will really help get your colonies off to a strong start.

Remember that since honey bees will not do what we tell them to do, then we have to arrange situations where we can direct their natural behavior. If you do this, then both you and the bees will have a more enjoyable time together.

SURPLUS HONEY

Question - I am a novice beekeeper (one year) with a question that has probably been asked a thousand times, but one that I need answered. How does one go about getting his tiny lady friends to put honey into a super?

I put up my first hive last April, and fed the bees until the end of May or early June. This is because we had a damp spring here in the Peninsula area of Washington. After I stopped feeding, the bees seemed to absolutely boom in numbers, and they were drawing comb at a terrific rate.

About the middle of July I checked the hive and found plenty of bees working, and decided to put in the excluder and add a super. The bees refused to enter the super, never drew any comb in it, and I finally took it off, figuring that I would give them the harvest of that year.

This year I would like to take some honey for myself, but I don't know how to go about it. Answers to my questions (from local beekeepers) have as many varia-tions as there are keepers, so I decided to ask the experts. Can you assist me? Any information that you might offer will be greatly appreciated. I read both the *American Bee Journal* and the "other" major beekeeping magazine every month. Thank you for your time.

Dan Brown
Port Angeles, WA

Answer - The criteria for getting a surplus of honey are bees, bees, bees and flower-ing nectar sources of abundance. You can have a high population of honey bees, but if there are no nectar sources, they cannot collect that which does not exist and vice versa. One of these criteria was not met either because after two brood chambers were filled with bees, perhaps they swarmed and reduced your population or disease or mites reduced the foragers' abilities or that there just were no flowering nectar sources.

It's hard diagnosing from long distance. I hope I've given you something to think about though. Better luck this year.

Question - Would it be all right to place my bees on the south side of my garage for the winter? They would have to be set right against the white building, so therefore would be warmed considerably on a sunny day. Would this variation in temperature decrease or increase their chances of wintering successfully? Would they require more than the usual amount of winter stores?

Milford Muggins
Munford, Massachusetts

Answer - You could probably winter your bees successfully on the south side of your white garage. It is important that the bees get warm at intervals to move their cluster and take a cleansing flight. It is also possible they would consume more food, but then again, they may rear more brood in the early spring which should make them a better colony.

BROOD REARING SPACE

Question - Since I began hobby beekeeping four or five years ago, I have been an avid reader of books and journals on all aspects of this endeavor. Not being an exact science, of course, numerous discrepancies between various authors are noticed. One, however, has troubled me the most. This concerns whether one or two deep brood chambers should be used.

Dan Hendricks
Mercer Island, Wash.

Answer - Studies have indicated that one deep chamber provides adequate space for one queen to lay and for brood to be raised in a standard honey bee colony. You are right that if there is not enough honey "storage" space that nectar being brought in will be stored wherever there is room - whether in the brood chamber or not. Many years ago research was done at the bee lab in Baton Rouge, La., that showed that instead of adding supers as needed that if the beekeeper would add in early spring 2, 3, or 4 supers all at once, the colony would actually collect and store more honey. The colony would "sense" that storage room was available and would try to fill it up. When a colony can store nectar other than in the brood chamber, it will.

So, yes, one brood chamber is fine in the summer as long as enough storage space is available. However, as far north as you live, we would recommend a second or even a third brood chamber that will be left on the hive full of honey for winter stores. Supers, of course, could also be left on.

Question - In your Dec. "Classroom" issue you made the statement that bees need a cleansing flight. I have felt that statement true. But then I read that someone is wintering bees in a dark cellar - one big long night 90 to 100 days long. Would like your comment on this. Also I want to comment on wax moth.

I have heard that cold weather like 0-Degrees kills the wax moth. It does not kill the larvae because when they hit the fire, they are flipping 1/2" high and are very much alive!

Harold Dill
Parker, CO

Answer - *As you know, the reason bees need to void, at an optimum no longer than every 30 days or so, is because their intestines are full of indigestible material. This material comes primarily from a whole variety of indigestible material naturally found in honey. The darker this honey, the more indigestible material is in it.*

Beekeepers who currently overwinter in climate-controlled buildings are striving to balance the bees' biological needs with their desire for low-cost efficiency. They try to reach this efficiency by overwintering smaller groups of bees, controlling the type of food they consume, the temperature of their environment, and its oxygen, carbon dioxide and humidity. In this climate-controlled environment, the temperature is kept at about 47ºF, the humidity is kept low, the carbon dioxide is monitored so as not to build up too high, and, as important as any item, all the honey has been previously removed from these small colonies. It has been removed so that it can be replaced with High Fructose Corn Syrup (HFCS). This is done because HFCS is a virtually pure sugar solution that has no extraneous indigestible particulate matter in it. As this small colony feeds from the HFCS, the need for defecation becomes almost zero because there is nothing to void. So if they don't have to defecate and they are calm and cool, you can keep them like this for extended periods of time.

You are correct that wax moth larva are overwintering in hives. Not enough research has been done to find out just exactly how they are doing it. The thought is that the larvae at this stage are very small and are keeping as close to the cluster as possible to stay above freezing. They last through the winter and are ready to start destroying any colony that has died over winter. It's nature's way of getting rid of disease in dead colonies.

TRANSFERRING BEES

Question - I have a hive that was given to me and the brood frames are falling apart from age. I have been told to take the old hive body, turn it upside down on top of a new hive body with foundation and, as the bees emerge, the queen will not lay eggs in the cells because of their downward slant. Please advise.

Bob Bartlett
Broken Bow, OK

Answer - Turning the old hive body upside down might work well and it might be interesting to give it a try for the sake of experimentation. A better suggestion would be to remove the hive cover, set a new body with combs or frames and foundation on top of the old hive body and replace the cover. The bees will work up into the new part and then the old body can be removed.

SUMMER MANAGEMENT

Question - Once the Spring Management period has passed, what generally are the phases a model colony might go through during the remainder of the summer?

Answer - Model colonies are somewhat like model children – you may read and hear a good deal about them, but they always seem to belong to someone else while your own (bees or children) are muddling along somewhere about average, here and there a success with a sprinkling of problems in between.

No beekeeper has ever written in to ask why he was having so much success – so we'll outline a typical sequence of events occurring over late spring and the summer with suggestions to browse here and there in this book to see what problems other beekeepers have encountered and how those problems might be solved.

When the development of the colony starts with the beginning of brood rearing in late winter or early spring, with a vigorous young queen, an abundant supply of food, and sufficient comb area for brood rearing, the colony's peak of population may occur at the start of the honeyflow, and the ideal in management has been attained. It usually requires 8 to 10 weeks between the time that spring begins and the start of the honeyflow to bring about this favorable occurrence. In other words, a fortunate conjunction of the peak in colony population and the beginning of the main honeyflow makes honey production relatively easy. Either too long a period before the flow or too short a time makes beekeeping more difficult.

If the honeyflow occurs in less than 8 to 10 weeks from the beginning of spring, there is often too little time for the colony to reach proper strength without the most expert care. It requires prolific young queens, an ample supply of food, and effective intelligent management – plus ideal conditions of weather, an abundance of honey and pollen plants, and freedom from diseases – to produce a maximum honey crop. If conditions are not ideal, it may be necessary to bolster colonies with queen-less packages or to unite colonies prior to the honeyflow.

When the honeyflow occurs 12 weeks or more after the beginning of spring, the colony may have reached its peak of population before the start of the flow. It will swarm before the main honeyflow and decline in its honey-gathering ability. To delay the peak of population in a long period before the honeyflow, colonies may be divided during the time of early flow from dandelion, fruit bloom, or other sources. A new queen is given to the queenless part and each division is allowed to grow until the honeyflow begins. The two parts may then be united or may be operated individually if there has been time for each part to achieve a satisfactory population of field bees for the flow.

This period before honeyflow is also the time when many colonies follow a natural instinct to divide their populations through swarming. Because neither the parent colony nor the swarm will gather as much honey as the original colony, swarming should be prevented or controlled by proper management.

Assuming that your colony is entering the honeyflow with a strong population, has not swarmed, and is headed by a queen able to maintain the colony strength through the honeyflow period, there is little else to be done to improve the colony and it is ready for the honeyflow.

The remaining work for the summer is supering and maintaining the colony in good condition.

SPRING MANAGEMENT

Question - I will be starting four new hives this spring. They will have all new foundation and new 3 lb. packages with queens. What distance do you recommend between hives? Is it all right to have several breeds of bees at the same place in different hives? How often do you recommend changing old foundations with new ones? I've read articles both pro and con concerning queen excluders. Do you recommend them? Finally, my inner covers have one side with very little space and one side with more space. Which side should face the bees?

Howard Wagner
Girard, PA

Answer - 1) Hives can be right next to each other, but it is best not to put them in a row. In a row the bees have a tendency to drop their loads of nectar and honey off at one of the colonies at the end of the row and not fly to their own colony. Either put them in a circle with the entrances facing out or in a four colony square with entrances at 90º angles to the entrance of the adjacent hive.

2) Having different breeds or strains of honey bees next to each other isn't a problem.

3) Foundation used for brood rearing should probably be completely changed every five years. This will remove a disease reservoir from the hive. Start a program of comb rotation and replacement where some damaged and old brood combs can be replaced every spring.

4) I use queen excluders successfully, but make a change in entrance loca-tion. I did some research many years ago at OSU/ATI showing that if you give an entrance above a queen excluder, your honey yield will increase.

5) Inner covers were designed with a shallow and a deep side – the deep side up for winter and the shallow side up for summer use. I don't believe the bees care or benefit that much from one side or the other. The deep side up in the winter pro-vides space for feeding granulated sugar or bee candy. The shallow side up in the summer gives the bees slightly more open space under the inner cover since bees populations will be larger and more nectar is being collected.

Question - Is it okay to put two hives side by side of different races of honey bees such as Caucasian and Midnite?

David Beavers
Orlando, FL

Answer - Different races of honey bees live quite well next to each other. There may be some drifting of bees from one hive to another. This is to be expected and will cause no harm.

Question - I've been keeping bees going on three years now. I have read all the bee-keeping material I can get my hands on, yet I always manage to notice something that I haven't caught in the books and magazines.

I overwinter my hives using two deep supers for each colony. Upon opening two of my strongest hives last spring for inspection, I noticed that between the bot-tom of the top super and the top of the bottom super the bees had built comb and there was considerable amounts of drone larvae being raised there. I couldn't miss this because when I pried open the hives the comb was torn in two exposing the glis-tening white larvae.

I suspect that I or my equipment has caused a clear violation of The Rev. Langstroth's "Law of bee space" and with the spring buildup in full swing, the bees were just making good use of this space.

What's your opinion and advice? I would like to avoid this problem in the future. I'm forced to scrape the comb and larvae off wasting my time and energy, not to mention the bees' time and energy.

K.A. Wareham
Moses Lake, WA

Answer - You are absolutely right, the dimensions between frames in the two brood chambers are incorrect. As you have probably figured out, it is either the width of your frames, the length of the rabbet (the little shelf the frames hang on) or the width of the box.

There is really no standard which each beekeeping woodenware manufactur-er adheres to in order to make all woodenware compatible regardless of who made it.

If you are only into the brood chamber area a couple times a year, you may be able to put up with the burr comb problem. If not, you are going to have to trim the width of your boxes down in order to establish a proper bee space.

There is not a whole lot of choice here. Either you put up with it or get your tape measure and saw out.

Bees are always going to find a way to build some burr comb and rear drone brood during the big spring buildup. This is not always looked upon as a problem by all beekeepers since it is a good indication that the colony is queenright, strong and healthy. The extra drone brood also provides a good source of brood to check for varroa mite populations without scraping out drone brood from the combs themselves.

Question - 1) How often should I cull/replace brood combs to help a hive from producing "Too small" bees?

2) Is there any treatment or wash that could be used periodically and satisfactorily to dissolve and clean-out pupal "castings"/casings from old brood comb, thus making combs with larger diameter cells (Like new)?

3) Will such smaller bees be more subject to a mite kill? Tracheal, varroa, or either?

4) If "Drone Foundation" produces drone cells, and "worker foundation" produces worker cells – It is possible to have bees draw foundation in between those two sizes, and produce worker bees in that comb? (Assuming, of course, that a larger worker bee might be advantageous)

5) What is the possible inside diameter size and tolerance of a worker brood cell that can normally be found in the field?

6) How harmful is it to feed medicated syrup to bees when they can't fly (due to cold weather)? Or, will the feeding of medicated (fumagillan and Terramycin) syrup kill a colony when they can't fly it off?

I ask the question #6, because I am looking for answer as to why I have lost so many colonies in the last three years 73% (11), 65% (10), and 86% (13) – even though I have alas, medicated for mites into the cold weather.

Roger Raymond
Southport, CT

Answer - 1) What's Too Small? Yes, there some correlation between bee size and the important tongue length, honey stomach size and wing size. It would probably take 15 to 20 years for larval skins to build up to the depth that would contribute significantly to downsizing the bee. You may want to cull every 5 years for disease reservoir reasons rather than size reasons.

2) Tough to do since everything is coated with propolis and wax. I do not currently know of any. A product called "Comb Cleaner" was once marketed, but did not prove itself in the marketplace where it counts. The product produced a harmless mold that the bees readily removed (and with it, hopefully, the larval skins also).

3) There is some minor debate that smaller bees develop more quickly from egg to adult and thereby disrupt the varroa life cycle and that smaller trachea on smaller bees inhibit tracheal mite entrance. Quite a lot of "small" cell foundation has been sold based on this belief. Personally, I'm not convinced.

4) Yes. In fact, I just read a Master's thesis from 1931 by Roy Grout of Dadant & Sons, Inc., that seemed to indicate that a larger bee would have a longer tongue, bigger honey stomach, etc., etc.

5) I do not know right offhand and truthfully haven't even looked. But there are approximately 857 cells per sq. decimeter for European worker bees and 1000 for Africanized honey bees as a comparison.

6) I think you should start feeding in fall when they can still get out and void, as these medications "clean" them out. Feeding medicated syrup when voiding is not possible, I think, would be detrimental, but have not research to back that up.

There is a new theory making the rounds that both mites are acting as vectors for a harmful virus or virus-like particle and that is what is causing so much loss in otherwise well taken care of bees.

Question - A thought came to me about honey bees and their watering problems. As long as they can get good clean water that isn't stagnated, they would be in good shape. But suppose they run out, cannot get water that is clean, where and what do they do?

Raymond H. Irwin
Woodward, OK

Answer - *The old adage "never assume" applies here. Don't assume honey bees like or better yet can find clean water.*

Water is a commodity like nectar that is necessary for hive health and air conditioning. When honey bees find a source of nectar, they get a sample of it, fly back to the <u>dark inside</u> *of the hive, give out sample for taste, sugar concentration, odor, etc., and do a "dance" to give directions to where it can be found. Nectar has a taste and odor and a sugar concentration that bees can identify. Clean, fresh, cool, sparkling water sources have very little taste or odor. But muddy puddles, ditch water, swimming pools have a lot of flavor which is necessary for bees to find it. If the honey bee coming back with water sample has one that is fresh and clean with little identifying taste, it is very difficult to direct the other bees to a specific spot. If the sample they received in the hive doesn't have a taste, they have lost this cue to match up a similar taste.*

Let's try this analogy. Pretend you have a terrible cold. Your taste buds have ceased to function normally and your nose is so plugged you can't smell anything either. I put you in a completely dark room and give you a bite of Brussel Sprouts. Remember you can't see the Brussel Sprouts and because of your cold, you cannot taste or smell them. (which in the case of Brussel Sprouts is probably a plus) Then, I tell you to walk out of the room, go outside and walk approximately 3 blocks north, and 2 blocks east and I want you to find and bring back the stuff I just gave to you. Your reply would probably be I can't because I don't know what it is you want me to get. Well, it's the same with honey bees – if they have a taste, odor clue like muddy ditch water, they have something to compare the sample with and match it up with the source. If they don't, they won't. Yes, clean water is better for a variety of reasons, but it's hard to find so they go with the easiest, the stinky water. Having a pond or stream near your beeyard is nice because it presents a nearby and relatively clean source of water.

CHAPTER 4

SWARMING

Question - I have a question on the results from a swarm prevention method. In the May 1996 issues of ABJ pg. 347-348, the author talked about different ways of preventing swarms. One method was to physically switch hives in a beeyard. A hive containing a strong colony would be switched with a weak colony to prevent swarming of the strong colony due to insufficient brood nest area. The workers will thus return to the location that they had memorized as their own colony when in actuality, they are returning to a different colony. What I wonder about is how do the guard bees react to a large number of returning workers which do not contain the colony's scent? The weak colony's guards more than likely would be flooded by the number of workers (formerly of the strong colony) returning with nectar. These bees will be able to take on the new colony's scent. However, the strong colony's guards should be able to rebuke the small number of returning weak colony workers. These workers, not being able to gain access to the stronger colony's hive, will not receive the new scent marker and that generation will die without being accepted into a colony. May I have your opinion on my thinking?

Doug Johnson
Indianapolis, IN

Answer - I may be reaching a little with the following analogy, but here it goes. Let's say you are sitting at home one day and you hear a car drive into your driveway, car doors slam and you hear footsteps coming up your walk to your door. You get up to see who is coming and it's two ladies in business attire. They ring your bell, you open the door and ask them what they want. They say they are from the State of Indiana and that they have a refund check for you, may they come in. Without too much hesitation you probably will let them in to receive your check.

The same kind of things happens with honey bees. If workers arrive at the entrance with either nectar or pollen, they are let in. It doesn't matter where they come from, how old they are, what color they are or anything else. As long as they are bringing food and want to unload it in that particular hive, the guard bees could care less. It's free food, let's take it! They come in and unload their nectar or pollen and they can easily acquire the specific pheromones for that colony which allows them to stay long-term in that colony. "Drifting" from one hive to another is a common occurrence and is the reason beekeepers are cautioned not to place their hives in long neat rows. The end colonies become the strongest because of drift.

If a worker arrives at the entrance without a food gift and does not have the colony pheromone signature, then she may not be allowed to enter.

Question - I live in the northwestern part of Pennsylvania and have kept bees, as a hobby for two years now. I've gotten a lot of information on beekeeping from various books, our local county extension office and the *American Bee Journal.* Nothing I've read has prepared me for what took place in the past two weeks. Two of my three hives have swarmed. The first and largest swarmed on August 21, right at the beginning of the goldenrod/aster season and the second swarmed on September 4. Fortunately, I was able to capture both swarms.

I've read numerous articles on swarm prevention, but they all referred to May/June swarming. This one caught me off guard. Both hives had new, well laying queens, the honey supers were only about half full, and there didn't seem to be any indication of overcrowding, temperatures in the 70's and 80's.

Is swarming unusual this late in the season? I check my hives about every one and half days to two weeks. Is this too often?

Bill Trumbull
Ridgeway, PA

Answer - *Swarming as late in the season as you report is unusual. But, I have received more reports the last couple of years that this activity has noticeably increased. There have been two prime explanations given for this late swarming activity.*

1) In many areas of the country, weather patterns have been cooler and rainier than normal. The thought is that since swarming occurs within a few weeks after nectar and pollen sources are abundant and following an over-wintering or prolonged dearth period, that weather may be involved. In many parts of the Midwest and East the nectar-producing plants were slowed, delayed or nonexistent during most of the cool, rainy summer. When the dependable goldenrod and aster bloomed and the bees could finally get out, this environmental/biological chain of events was put into action with the result being more late-season swarms.

2) Many times when environmental, disease, or parasite problems occur, honey bees will go into a genetically programmed survival mode and do whatever they can to leave a bad area or situation behind (absconding). With the advent of mites, tracheal and varroa everywhere, more swarming has been seen at historically non-swarming times. Honey bees in Africa and the tropics sometimes react to high mite infestations (varroa) by breaking the brood cycle and swarming. In effect, leaving behind the sick, the dying and most of the parasites. Our European stock of honey bee does not have a high degree of this absconding behavior, but it does exist in the population and is exhibited at times.

3) Your third choice is that your experience was just one of those quirky things that happen occasionally.

It sounds as though you are doing everything right and are a conscientious beekeeper. Keep it up.

SWARMING

Question - Where is the queen when the bees swarm into a tree and how long will they stay before leaving?

Answer - When a swarm clusters away from the hive, for example, on a limb of a tree, the queen goes with the swarm – unless clipped or injured. A number of times when we have attempted to find the queen in a swarm, it seems that she would be near the top and towards the center of the cluster of bees.

Once the swarm has clustered on a tree or such, it may remain there from 15 minutes to an hour or even a whole day. Sometimes when the weather is bad, the swarm may hang there several days. But the normal condition is for the swarm to break up and move on to their new home within a few hours of having settled.

There are times when a swarm may, for some reason or other, decide to stay in its position and they will start to build combs. The queen will start laying in the combs, and soon you have an outdoor swarm. This probably happens when the old queen which left with the swarm was still rather heavy from egg laying and finds it difficult to fly. Another factor which may affect this is the honeyflow. If it is quite heavy, then the bees find it easier to produce wax and build the new combs necessary for this type of colony.

YOUNG QUEENS FOR SWARM PREVENTION

Question - I am told that I can prevent swarming by keeping my colonies stocked with young queens. Is this reliable?

Answer - There are many causes of swarming: overcrowding, improper ventilation, and supersedure impulses to name a few. Supersedure of the queen is Nature's way of replacing a failing queen and by keeping your colonies stocked with young queens, you will have forestalled one of the swarming impulses. However, you will still have possibilities of swarming from the other causes unless your management program also solves these problems before they reach the point of swarming.

Since requeening with young queens alone will not solve the entire swarming problem, keep a close watch on all of the potential problems a hive might have during a working season and requeen only when it becomes necessary. Many queens will give good service for two years.

DRONE EGGS IN SWARM CELL CUPS

Question - Does a queen ever lay drone eggs in those small swarm cell cups at the bottom edge of the frames or must one always assume that when they have eggs in them that queens are being raised?

Answer - Under normal circumstances, we doubt very much that the queen would ever lay drone eggs into the small cell cups which are frequently found on the bottoms or ends of the combs. We would assume that when a normal queen lays eggs in these cups, that the colony is preparing to swarm or, in the case of a failing queen, preparing to raise a supersedure queen. Under abnormal conditions, such as an old queen which has no more sperm left, an unmated virgin queen which has started to lay, or in the case of laying workers, it might be possible for a queen to lay drone eggs in the cell cups.

CLIPPED QUEENS AND SWARMS

Question - What happens to a clipped queen when she tries to swarm and must return to a hive full of virgins?

Answer - Under normal conditions, when a colony tries to swarm with a clipped queen, the bees would stay with the queen even though she was able to get only a short distance away from the hive. We have noticed many times that the swarm will leave the colony and fly to a nearby limb to light in a normal manner. However, they soon realize that the queen has not followed them to this limb and become quite nervous, eventually returning to the hive. If the queen is on the ground nearby, there is usually a small group of bees with her. We have returned the queen to the hive after cutting the queen cells and killing any virgins which might be out. If, however, the bees tried to swarm a couple of times like this and failed, the next step for them would be to swarm with one or more virgins.

VIRGIN QUEEN FLYING OUT WITH A SWARM

Question - Will an unmated queen fly out with a swarm?

Answer - A virgin queen will fly out with a swarm, especially if the queen of the parent colony is clipped and cannot fly. Under such circumstances, a virgin queen may join the swarm. At other instances, virgin queens may fly out to mate and they may be joined by small swarms. These usually return to the parent colony after or before the virgin queen has mated.

G. H. Cale, Sr. states in The Hive and the Honey Bee that in very populous colonies, about a week after the prime swarm leaves, more bees may swarm out with virgin queens, apparently when the virgins take their mating flights. There may be several virgins in each of these swarms and this type of swarming may continue until the population of the colony is reduced to a low point.

SIGNS OF SWARMING

Question - Is there any way I can tell by the bees' actions when they want to start swarming?

Answer - An experienced beekeeper could sometimes tell by the action of the bees whether or not the colony is likely to swarm. Usually there is more drone activity, the bees seem to fly a little less and act more lazy. Hanging out in front of the hive may be a sign of impending swarming, but it also indicated other conditions such as overcrowding, lack of ventilation, or hot, humid weather. To be on the safe side, check the colony every eight days or so for queen cells.

PREVENTION OF SWARMING

Question - I am a retiree and plan on keeping a few hives of bees. Would you please tell me how to prevent them from swarming and also what to do when they are about to swarm? Would you also tell me what kind of bees to start with?

Answer - I'm afraid that no one has the answer as to how to prevent all swarming. The following conditions seem to contribute to swarming, so if you correct these conditions, you should have the minimum amount: (1) poor queens result in supersedure; (2) crowded conditions for brood and storage of honey; (3) poor combs with too much drone comb; (4) colonies which do not go into the supers fast enough.

What can you do? Be sure the colonies are strong enough to go into the supers. Requeen every year or two, or whenever you find a failing queen. Add supers before the bees need them, rather than after. Don't use a queen excluder until after the bees are well established in the supers. You will have fewer problems with extracted honey production than with comb honey. Use only good combs so that the amount of drone brood, which is of little value to your colonies, is at a minimum.

Most beekeepers use Italian bees but this is a personal thing. The Caucasians have some fine points and are worth a try. Why not try both Dadant Starline and Midnite hybrids and see which you like better.

INSPECTING COLONIES FOR SWARM CELLS

Question - How often should a colony be inspected for swarm cells? Would it be wise to destroy drone cells when it is possible as a swarm control?

Answer - A colony should be checked for queen cells every 8 days during the period before the honeyflow or the swarming season. It is a waste of time to destroy drone cells. You could eliminate the problem of drones by using a good queen and good worker combs.

QUEEN EXCLUDERS TO PREVENT SWARMING

Question - I was wondering why beekeepers don't simply put on a queen excluder on the bottom of the hive to prevent swarming by preventing the queen from leaving the hive.

Answer - The method which you described in your question is used sometimes by beekeepers to prevent swarming. However, we do not advise using a queen excluder for this purpose for several reasons.

First of all, the queen excluder can become clogged with dead drones and inhibit the entrance of the worker bees. Also, even if the excluder is kept clean, it still does provide an obstruction which the bees must pass through and thereby could lower your honey yield. Sometimes an abnormally small virgin queen will be able to slip through the queen excluder and swarm anyway.

We think it would be much better for beekeepers to take preventive measures by providing the colony with enough room during the honeyflow to prevent them from swarming. Even if a queen excluder is used, this will not prevent the colony from building queen cells during crowded conditions.

THE ALLEN METHOD OF SWARM PREVENTION

Question - What is the Allen method for swarm prevention?

Answer - We went back to the January, 1960 issue of ABJ to let Dr. C. C. Miller answer your question: "When the honeyflow is well started, I go to each strong colony, regardless of whether the bees desire to swarm or not, and remove it from its stand, putting in its place a hive filled with empty combs, less one of the center ones. Next, a comb containing a patch of unsealed brood about as large as the hand, is selected from the colony and placed in the vacant place in the new hive; a queen excluder is put on this lower story, and above this a super of empty combs, this one having an escape hole for drones; and on top of all, an empty super. A cloth is then nicely placed in front of the new hive, on which the bees and queen are shaken from the parent hive, and the third story is filled with the combs of sealed brood and brood too old to produce queens, and allowed to remain there and hatch, returning to the working force.

"This is really the Demaree plan, which was given to the public many years ago by G. W. Demaree, a prominent Kentucky beekeeper at that time. Mr. Allen has varied it by putting a frame with some brood in the lower story, whereas Mr. Demaree had only empty combs or combs with starters in the lower story."

DEMAREE METHOD OF SWARM CONTROL

Question - What is the Demaree method of swarm control?

Answer - The brood of the colony is examined and all queen cells are destroyed. The hive is then removed from its bottom board and a body containing one comb of unsealed brood, eggs and the queen is put in its place with the remaining space filled with empty combs. A queen excluder and the remainder of the brood and the bees are placed at the very top. The colony still has all of its brood, and the queen is in the lower body with a free brood nest.

In 10 days, examine the brood combs in the top hive body and remove all queen cells that may have been built in the interval. In 21 days, all of the brood will have emerged in the upper body and it will be used for honey storage, while bees will be beginning to emerge from new brood in the lower body, so that a continuous succession of young bees is maintained. Except in unusual seasons, it is seldom necessary to use the Demaree plan more than once.

THE TWO-QUEEN SYSTEM AND SWARMING

Question - I have been using the two-queen system for several years. Last year nearly all the colonies swarmed. All but one or two of the old queens got lost when swarming and the swarms went out again with virgin queens. How can I improve on this situation?

Answer - It may be that the two units were united too early. They should not be united until the flow has developed to the point where each single unit is well started in nectar collection. If united before this, there are too many bees with too little to do, so swarm cells are started.

Assuming that your swarming occurred in springtime, it may have been due to inclement and intermittent weather which confined the bees to the hive and in such circumstances they need something to do such as drawing comb foundation. Some beekeepers make a practice of giving supers of comb foundation to draw the worker bees out of the brood nest and to give them something to do. However, it sometimes seems that the bees do not read the same instruction books as we do and there are times when weather and other conditions result in their swarming regardless of what we do. There also some years when swarming is worse than in others.

SWARMING

Question - Last week I found the queen (clipped and marked) on the ground with about 50 bees around her (not a swarm). I picked her up, put her on the entrance and she went in. I then put on a deep hive body for more room and extra food supply for next winter. I understand many in this area use the two deep hives for brood rearing and then later add shallow supers with a queen excluder. Why did the queen leave the hive? Also, when should I put the original hive brood chamber which is about full on top of the one I just added?

Answer - It does sound as if the bees attempted to swarm with the queen, but inasmuch as she was clipped, she could not fly with them. You did the proper thing by placing her back into the hive, but you might have done well to check to make sure there weren't any ripe queen cells in the unit. If there were and the old queen was not allowed to get at the new virgin, the hive could still swarm with the young virgin. Incidentally, one deep hive body is probably not sufficient to handle the brood rearing. We would recommend that two deep hive bodies be used for the brood chamber and then your honey supers placed above that. This gives more versatility for you in the spring as far as manipulating the hive to stimulate buildup, and, as you mentioned, gives more food storage space for the winter.

The normal sequence of events in the hive is for the bees and queen to move upward in the hive with the egg laying as the queen needs the space. Usually, when the honeyflow begins to come on strong, the worker bees will force the queen back down into the brood chamber again. However, in earlier spring she may go up through the entire hive, leaving the bottom completely empty. In advance of this happening, many commercial beekeepers practice what is called hive reversal. Both bodies are removed, the top one placed on the bottom board and the other on top. Then about two weeks later, they'll go back and re-reverse as necessary. This keeps plenty of egg laying room above the queen's head without letting her get clear up to the top of the colony.

PREPARING FOR A SWARM

Question - Last spring I acquired a swarm of bees from a friend. Due to the lack of equipment and a lack of know-how on my part, they swarmed again before the end of the season. I happened to be on hand at the time and it was a beautiful sight.

In anticipation of a swarm this season, I would like to be ready with a prepared hive. How may frames of foundation should I have in the hive.

Also, about 1/4 mile away is a bee tree (probably my swarm), too inaccessible to capture. From a disease point of view, should they be destroyed and what should be used?

Answer - It would be an excellent idea to get a hive prepared ahead of time. We assume that you are using the standard 10-frame hive. If the swarm that you catch this spring should be a large one however, it would be fine to use a full set of frames and foundation in the hive. It would be even better if you had some frames of drawn comb.

It is doubtful that there is any need to bother the swarm in the bee tree. Should it become diseased, the colony will probably die. Should this happen, then it might be wise to close the hole which the bees are using as an entrance.

DECOY HIVES FOR SWARMS

Question - In late June I prepared an empty hive quite some distance from the two hives I have, in anticipation of a swarm. The swarming took place at a time when I was away. And to my surprise I found them a few weeks later in the hive I had prepared. The shallow super that I had placed on the hive was full of comb and was about two thirds full of honey. Was this just luck or did I do something that encouraged them to do this?

Answer - It is almost like E.S.P. in that they knew what you were thinking or planning on. This is a purely natural response from a swarm of bees to move into a hive like this. In fact, many beekeepers frequently leave "decoy" hives in or around a bee yard so that during the season stray swarms may build in them. We have also heard of using a small ball of cotton placed in the back of the hive with a little anise oil sprinkled on it to attract the bees. This is something that you may wish to do again in other years.

NOISE TO CAPTURE SWARMS

Question - I have read of bees calmed with sound one-half octave higher than the middle C note on a piano. I have also heard old beekeepers tell of bringing down a swarm by pounding on a dish pan. Maybe if they had the right size pan that gave the right vibrations per second it would work. What would you say to that?

Answer - We understand that some work has been done on bees' reaction to sound, but don't believe it had any practical value.

We have also heard of swarms being brought down when the beekeeper beats on a dish pan, but have never put much faith in the idea. The beating on a dish pan, as far as we know, came about as a result of an old English law which stated that the beekeeper must let everybody know that he is following a swarm to keep it in his possession. We would guess that beating a dish pan would serve to attract the required attention, but we have no verification that it had any effect on the swarm.

REMOVING SWARMS FROM BUILDINGS

Question - I am writing regarding two swarms of bees in a building. They are about six feet from the ground. What would you suggest to remove them? I was thinking of covering all the outlets but one and fix a wire screen with a bee escape to the new hive. Now the question. What do I do from here? What about a queen? These bees have been in the building for years.

Answer - Your idea will work, but you must either put a queen, queen cells, or a comb of eggs in the hive next to the bee escape so the bees will be able to rear a queen. It takes about 30 days to get all the bees out of the building. It is possible to then have the bees rob the honey out of the house.

CATCHING SWARMS

Question - I was always wanting bees. At last I bought two colonies which will swarm and I don't know how you catch the swarms. I wonder if you will help me by telling me how it is done? I'm bothered about it because I don't know how or when they will swarm. In fact, I don't know anything about bees but I am very interested.

Answer - Books have been written on swarming and its control; so you see it would be difficult to explain in one letter. We suggest that you do purchase some good books on beekeeping such as The Hive and the Honey Bee and First Lessons in Beekeeping and that you do some reading.

Bees usually swarm before the main honeyflow. In this area, we expect most of our problems in May and early June. Queen cells will appear in the brood chamber to indicate their intention to swarm. The bees (about half of them), the old queen, and some drones will leave just before the new queen emerges. When they cluster on a limb or brush, it is fairly simple to hive them. place the hive under the swarm and shake the bees so the bees fall in front of the hive. They should walk into the hive without coaxing. Some old combs in the hive will make it more attractive.

Some swarming can be prevented by keeping the colony headed by a good young queen, giving the bees plenty of space for brood rearing and for storing honey. Partial shade and good ventilation may also be of some value.

HIVING A SWARM

Question - In August I found a small swarm in one of my yards. I killed the queen and they went back in. Since then I haven't been able to find a queen or any sign of one. I put in a frame with eggs and brood but they didn't start cells. I tried to introduce a queen but they killed her. In October when I covered them for winter, they still had quite a few drones. Could they have a virgin and are keeping the drones to mate with her next spring? Do you think killing the queen and letting them go back to the parent colony would be all right?

Answer - The colony in question either has a queen or will never have one. Usually a colony which has a queen will neither accept another queen nor will it produce queen cells even though you put in a frame of eggs. Once in awhile we find colonies which will not accept a queen by any of the usual methods. In this case the best procedure is to place the colony on top of a queenright colony with a double screen or sheet of newspaper between the two colonies. The bees will chew through the newspaper to become united. The double screen should be removed, if used, in a week or so to allow the colonies to unite.

The presence of drones could mean that the colony has a virgin, but it could also mean the colony is queenless. It's perfectly all right to kill the queen in the swarm and let the swarm go back to the parent colony, but you should try to find out why the colony swarmed and correct the condition. Some beekeepers make it a practice to requeen all swarms to try to improve the stock.

CROSS COMBS FROM A SWARM

Question - I have a colony of bees which are established from a swarm. The bees were dumped into a hive without any foundation and left to shift for themselves.

They built cross combs. Last fall after the main honeyflow, I put on a deep super of wired foundation. This spring I found the bees had drawn the foundation into combs. I then put the top super on the bottom and the bottom on the top in hopes that the bees would move down but they wouldn't. How could I move the bees down so I could remove the top super?

Answer - If you allow the bees to become crowded, the queen will probably go into the bottom hive body with the good combs. You can then place a queen excluder between the two bodies and take off the crossed comb super in about a month, after all the young bees have emerged. We usually do it the other way; we place the good combs full of honey on top for the winter. When the bees are found in the hive body in the early spring, the crossed combs are removed. It is possible to smoke the bees out of the crossed comb into the one with the good combs.

FAILURE TO WORK IN THE SUPER

Question - Last year we hived a swarm of bees and gave them one super. When we examined the colony in late fall the bees had not stored any honey in the super although the lower body was full of honey. Can you tell me why they did not work in the super?

Answer - Perhaps there was not sufficient nectar available for your bees to store any more honey than they did. Also, bees do not like to go up into a super of foundation. It is often necessary to move some frames of brood up into the super from the brood chamber, providing the super is of standard depth. The frames of brood in the super entice the bees to move up and go to work. You might have an old queen in this colony if it has not been requeened since the swarm was hived. If so, she is not raising enough brood to make a populous colony with the good field force for bringing in nectar. Your bees should be inspected often and in swarming season, queen cells should be cut out. Try requeening your hive and see if that helps.

HIVING A SWARM

Question - I recently captured a swarm of bees and placed them in a new hive with proper frames and foundations. My questions are: (1) After three days I inspected the hive to find most of the bees clinging in a swarm on the inside cover. Does this indicate that I have possibly lost the queen? (2) Is it true that the queen with a swarm is an old queen and should be replaced? (3) Is it possible to save this swarm if it is queenless by introducing it into another hive of bees that does have a queen or should I requeen?

Answer - In regard to your first question, we suggest that you bounce the bees into the body and close the hole in the inner cover. You should also check the colony and make sure that the queen is still alive and in good condition. If you cannot find her or any evidence that she has been laying, then you should immediately obtain a new queen for the colony. This time we would suggest that you have her clipped and marked if you have not done so in the past. This facilitates the finding of her on the combs.

Yes, you are right in your belief that the laying queen is usually the one that goes with the first swarm. Subsequent swarms, if any, will have virgin queens. However, it is not necessarily so that the queen is past her prime for egg laying. Therefore, you should wait and see how she lays before you think about replacing her.

In regard to your last question, we would suggest that if the swarm is rather small that you go ahead and unite it with one of your other colonies. However, if the swarm is rather large and in good condition, we would suggest that you requeen the swarm to save it.

MAKING A SWARM STAY IN A HIVE

Question - How do you make bees stay in a hive when you hive them?

Answer - We assume that you are referring to catching a swarm. Sometimes it is difficult to keep the bees in the hive if all of the frames have foundation in them rather than combs. If, when the swarm is being put into the hive, the queen can be found and one wing clipped, this will prevent her from flying away again. Sometimes the queen will leave the new hive and when she does, the colony of bees goes with her. One thing which you can do to help keep the bees in the hive is to make certain that there is a feeder full of sugar syrup at the time the bees are hived. This will keep them from being hungry. Many times a colony of bees will leave if they are in a starving condition.

EXCESSIVE SWARMING

Question - I reduced my colonies to 41 standard hives and last fall left one or two full supers on each. Also I have been requeening in July for several years with three-banded Italians. Then in March I had a large swarm and each day thereafter one to six swarms ending May 29th. In all, I hived 79 large swarms. Also 33 swarms left because I had no hives.

Why this unusual urge to swarm? How can I prevent it? Was it because of too much honey left in fall? Requeening too often?

Answer - You must have set some sort of record in collecting 112 large swarms from 41 colonies. Do you think that some of them might have been bees from other sources?

I believe much of this swarming could have been prevented by adding supers earlier. This is one of the problems which arises when so much honey is left with the colony over winter. One and a half or two-story colonies work well here, but we find some advantages in a little early feeding of sugar syrup to stimulate brood rearing.

If your colonies build up too quickly, you might divide the colonies to help reduce the swarming. Conditions might also have been unusual and the same problem may not come up again in years.

We doubt that either too much honey or too frequent requeening will cause the problem of excessive swarming. It would seem as though you had too many bees with too little to do and that lack of space in the hive was the main problem.

AFTER-SWARMS

Question - I am a beginner with only six colonies, two with Starline queens and four with dark Italian queens. At the beginning of last spring, I had two colonies, one with a Starline, the other dark Italian. The Starlines swarmed once, the swarm weighing about 9 pounds. The dark Italians swarmed three times, once in late April, again in early June and again July 15th. The swarms weighed 8, 10, and 15 pounds and yet the hive is still full of bees and emerging brood. Should I attempt to stop these after-swarms?

Answer - *You should try to stop not only the after-swarms, but also the prime swarm. It is best to keep bees from swarming because the queen, a large portion of the bees, and part of the honey goes with the swarm. Also, there is a period when the colony is left without a laying queen. Swarming may be controlled in part by early and adequate supers, good queens and combs.*

CHAPTER 5

MOVING BEES

MOVING BEES

Question - I tried to move my bees for the first time. What a mess. I put wire in the hole and nailed down the hives. I rented a cart to move the hives to the truck. The nails didn't hold the hives together. The wires came out and let the bees out. I let the bees calm down a little and tried again. I lit the smoker and blew smoke into the hive and tried to put the wire back in, but as soon as I would stop blowing smoke the bees would be out before I could put the wire in.

Answer - You have certainly had a lot of trouble with the hive that you tried to move which came apart. The most commonly used method of securing the bottom board to the hive is to use hive staples. These staples can be used to fasten the supers to the hive body or supers to supers. They are large enough that they will hold securely against most road bumps.

MOVING SCREENS FOR HIVE ENTRANCE

MOVING SCREENS ON HIVE

Question - I have enjoyed raising bees for about four years. I think I have done well, but I have made so many mistakes. But now, a new and serious problem has arisen, we also raise horses. I have a beautiful two-year-old colt.

For some reason the bees seem to pick on him and he goes crazy with only one bee. Recently I ended up with a dislocated shoulder and was taken to the hospital.

Also, lately, the bees have been attracted to my hair. Can hair spray affect them?

I love beekeeping and sell lots of good white honey. Can you help me overcome this problem so I will not have to give up beekeeping?

Answer - Your letter was quite interesting. We think it would be advisable to move your bees to a different location away from the horses. This might be an inconvenience but the risk warrants it. When stung, a horse will not usually run away. Instead, it will kick at the hive. This makes the bees all the more angry and they come pouring out of the hive ready to sting anything in sight. Horses have been seriously injured or even killed by numerous bee stings.

Although we have not recorded cases of bees being attracted to one's hair, it is quite possible. If the bees are angry, they are apt to go for a darker colored object such as one's dark hair. Whether bees are attracted to your hair spray or not is difficult to answer. This could be possible, however.

Question - I am in the process of relocating my beeyard approximately 350 feet from its present site because of it being better protected and beneficial to the bees. I have six colonies and my problem is (1) when is it best to relocate the hives to prevent drifting and also a minimum of loss of field bees, meaning at what time of the year and (2) how best to make the move? I am currently clearing the yard and plan seeding it with grass, shrubs, etc. and am using large cement blocks as hive stands and would like to make the move in the easiest manner possible.

Answer - If they were my bees, I would merely pick them up some evening and move them to the new location. Leave a weak colony behind to pick up the field bees which return. Putting grass or a sloping board over the entrance will help to make the bees realize some changes have been made. Move the last hive after another day or so. You should remove as many of the old landmarks as possible.

Many books suggest that you move them only a short distance each day. You will probably have fewer bees drifting with this system.

Question - I wish to move fifteen colonies of bees a distance of seventy-five to one hundred feet. This move is necessary because the owners of the land wish to drill test holes for fire clay where the hives are now located.

I have often moved beeyards a short distance during summer when bees are flying every day. I just simply pick up the three and four story hives with my wheelbarrow and move them about twenty-five feet each day. The bees seem to locate their hives after these short moves without much confusion. I can move a yard of fifteen three or four story colonies this short distance in about thirty minutes. It is necessary to make a number of trips to the yard to make the short moves.

I am wondering what would happen if I made the one hundred foot move at once during cold weather while the bees are not flying. Would they come out of the hives on the first warm day and return to their old location and be lost?

Answer - When we have moved bees in the past, for short distances, we have done as you have. That is to move in a series of steps. However, I can understand where a fifteen colony beeyard would present problems. You could probably make the move before the weather gets very warm in the spring. However, you can expect that some of the older bees, when they first fly out, will head back to the old location. You might consider putting up a "dummy" hive to catch the stragglers, so they don't become a nuisance for the people in the area.

**A LOAD OF BEES JUST DROPPED
OFF FOR ALMOND POLLINATION**

Question - I would like to know about moving bees. This spring I moved a number of hives to a farm about 50 miles from home. I put screen on the front of the hives to keep them in. This was quite a job because some of my bees are hybrids and they boil out with the first touch of their hive. The Caucasians are not like that and I believe I could move them back home this fall. Is there a better way?

Answer - Most beekeepers use a similar system of moving bees. We cut a screen the length of the opening and about 2-3 inches wide. This is pushed into the entrance in a "V" shape and because it is springy, it stays in place. This can be done before the bees boil out. We also smoke the entrance to the colony before placing the screen in the entrance.

Question - Can one move bees for a short distance or a long distance, say 15 miles when the temperature is 30º or colder – say to 10º below zero? If their entrance is screened while moving and after they are located, should this be removed or should one wait until they are settled if they are active?

Answer - You should have no trouble at all moving bees for the distance that you indicate in this cooler weather. Bees moved at these temperatures are not going to be active. Any that are shaken from the cluster will probably chill and die. I wouldn't think you would need to be concerned about having the entrance screened after they are placed in their new locations.

HIVES ON MIGRATORY PALLETS - READY TO MOVE

CHAPTER 6

EQUIPMENT

RADIAL EXTRACTORS

Question - I read in *Gleanings* that the parallel radial extractor is twice as fast as a normal radial extractor. Is this true? Why?

Mike Hunt
Los Angeles, CA

Answer - *Let's see how a radial extractor works. From page 38 of the 1992 Dadant catalog it says, "Radial extractors are the quickest, easiest and most efficient way to extract honey from the comb. Uncapped combs are placed in the circular reel like the spokes of a wheel," and spin around the vertical driveshaft of the reel. The reel spins creating the centrifugal force that forces the honey out of the cells. A regular radial extractor spins the frames around a vertical axis. A parallel radial extractor does the same thing only around a horizontal axis. In other words, the parallel radial extractor is just like a radial extractor except it has been turned on its side. The centrifugal force being created by the spinning action is the same no matter what direction or attitude a reel is in. Centrifugal force is centrifugal force. My guess is that the good people at Gleanings just published what they were given for an article because a parallel radial extractor will not, all things being equal, be twice as fast as a normal radial extractor in extracting honey.*

FULL LINE OF DADANT EXTRACTORS

Question - I am a hobby beekeeper and I'm considering building a brood chamber in the old "Jumbo" depth (11.25"). I am writing to find out if Plasticell foundation can be had in the depth. If not Plasticell, is another foundation available from you in this depth? If it is not available from you, do you know of someone else who produces and sells it?

Lew Wolfe
Portland, OR

Answer - Hope you enjoy your experimentation with the "Jumbo" size equipment. There is still some in use, but not many beekeepers appreciate the extra size and weight. Research reports seemed to indicate that honey bees did well in this size and may have been able to overwinter more successfully than in other sizes.

In any event, Dadant and Sons still makes crimp-wired foundations for the Jumbo frame with eleven sheets per box for $15.45 plus shipping.

HONEY PRESSES

Question - A few months ago I viewed a TV program about bees and the beekeeper used a "honey press" to extract liquid honey from the combs. I have been trying to locate one of these devices ever since. I thought you might have some ideas as to where I could obtain a honey press.

Raymond Williams
Binghampton, NY

Answer - Honey presses are needed and used primarily in England and Scotland where thixotropic honey from heather requires this method to extract it from the comb, unless special devices are used to disturb the honey in each cell enough to bring about some decrease in viscosity caused by a specific protein in the honey.

I have two addresses for you to try. They should be able to help you.

E. H. Thorne Ltd.	*Steele & Brodie*
Beehive Works	*Steven Drove, Houghton*
Wragby, Lincoln	*Stockbridge, Hampshire*
LN3 5LA ENGLAND	*ENGLAND SO20 6LP*

Question - Do you recommend using slatted racks as part of your bee hive setup? What are the advantages and disadvantages of using slatted racks?

Tim Armstrong
Ashtabula, OH

Answer - Slatted racks are one of those semi-interesting beekeeping appliances that someone asks about every year or so. Like everything else in beekeeping, you will find beekeepers who think they are great and others who say they are worthless and some in the middle.

If you are a hobby beekeeper, it is one of those things that is fun to play around with, but if you are trying to run your outfit on a profit basis, the slatted rack may just be another piece of equipment to keep up with that costs more money than it may give you in return.

For those of you who are new to the slatted rack, here is a little of what I know and what my friend Kim Flottum refreshed my memory on.

Back in the late 1800's there was a very well known, respected and if you will famous beekeeper by the name of C.C. Miller. Miller used a bottom board of his own design that was 2 inches deep. The standard bottom board today is generally 3/4 to 7/8 inch deep. Having a bottom 2 inches deep created some problems. The main-problem was the bees built comb from the bottom of the frames to the bottom board to fill up space. Mr. Miller, being the smart beekeeper he was, made a device using two rails running the length of the bottom board from the front to the back with slats nailed across the rails with 3/8" space between them. This device, when inserted in the 2" space in the bottom board, stopped the comb building and yet allowed bees to cluster in the space.

Anecdotal reports have said that the slatted rack relieves congestion and improves ventilation and thus helps prevent swarming.

I'm not aware of any "formal" research ever having been done on the slatted rack and its effect on a hive of honey bees. It's probably like many other things in beekeeping, sometimes it seems to work and sometimes it doesn't.

EXTRACTOR RPM's

Question - I have a two-frame extractor that I operated by hand until this year when I put a motor on it. The problem is that I do not know if it is revolving at the proper speed.

Shirley Clyde
Donnellson, IA

Answer - The extractor should operate at about 300 RPM. You should, of course, start at a lower speed so you can get a portion of the honey out of one side first, then reverse.

Question - I have seen the Strain Away Honey Filtration System advertised for sale in the United Kingdom. Is there any company in the United States that sells the system?

I am planning to build a honey house. I was wondering if I could line it with cedar wood. Would cedar wood be harmful to the bees?

Where can I get a book about how many hives it takes to pollinate crops per acre?

What about the cigar smoke in a hive? I lost 13 hives out of 18 hives this past winter and spring to varroa mites.

Bob Wilson
Abilene, Texas

Answer - *I have only seen the Strain Away Filtration System advertised in British publications. If the product is good enough, I'm sure there will be representatives in this and other countries eventually.*

A Honey House made of Cedar would certainly be pretty and would do no harm to the bees. If this structure will also be used for extraction and bottling, your ability to keep a clean and sanitary environment may be compromised by using any type of wood surfaces. Keep this in mind in your long-range planning.

On getting a good basis for pollination requirements I would start with Chapter 24 of The Hive and the Honey Bee written by Roger Hoopingarner and Gordon Waller. This chapter gives excellent information on crop variability and needs.

Cigar smoke, like any tobacco product, contains tars and nicotine which will kill or temporarily disable varroa mites. With the use of a sticky board, this is a good survey tool to see if Apistan Strips are needed immediately. Since tobacco smoke is also toxic to humans, we suggest you burn the tobacco in your smoker. That is unless, of course, you are a cigar smoker.

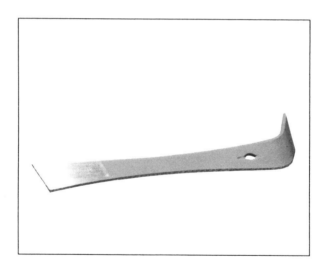

HIVE TOOL

Question - When I began keeping bees and reading bee magazines seven years ago "PEN-ETROL"® by Flood was frequently advertised as an approved wood preservative for use on hive components. Are they simply not advertising any more, or has approval been withdrawn? If so, are there other wood preservatives approved for use on BEE-WARE?

Based on my experience so far, all I'm really concerned with treating are my bottom boards and slated racks.

Walter Burrows
Exeter, RI

Answer - I had forgotten about "Penetrol". I even have 15 or 20 supers I treated with it 10 or 15 years ago. They are in good shape to this day, but I have never seen "Penetrol" advertised since nor have I seen it in the wood treatment aisle of our local Mega-Do-It-Yourself store.

Some beekeepers, primarily in New Zealand have large tanks where they melt beeswax and/or paraffin and dip their woodenware in it. The wood soaks up the wax and is preserved for a very long time.

Here is some info from The Hive and the Honey Bee on wood preservation.

PRESERVING WOODENWARE

Wooden boxes, tops and bottoms should be treated to maximize their life. Painting is traditionally the best preservation of the woodenware in beekeeping enterprises. Normally, supers are only painted with two coats on the outside. The inside is not painted to allow the wood to absorb moisture produced by a colony. Special attention should be given to the joints and exposed end grain.

Both water- and oil-base paints are effective. The former is used by most. Water-base paint is more permeable to moisture than the oil-base variety. Hive body color appears to be of little importance to bees and most are painted white as a tradition. The color reflects sunlight and may keep colonies cooler in summer. In the North, darker colors may be more beneficial to help colonies survive cold winters and increase foraging on cool days.

Many beekeepers rely on wood preservatives and do not paint their colonies at all. Research indicates that copper naphthenate, copper 8-quinolinolate and acid copper chromate are the best wood preservative options at present for beehives. Creosote, Pentachlorophenol (PCP), tributyl tin oxide (TBTO) and chromated copper arsenate (CCA) have adverse effects on bees and can be accumulated in wax (Kalnins & Destroy, 1984; Kalnins & Erickson, 1986). Most wood preservatives are classified as pesticides and must be used consistent with their label.

Constructing hives out of commercially treated lumber by the process known as "Wolmanizing" is not recommended. The process is variable and many different chemicals may be used. Evidence has accumulated that lumber can be made totally worthless for beehive construction when treated by Wolmanizing. Bees coming into contact with treated wood have died. It is also dangerous to humans; a sawdust mask is now recommended when working with treated lumber.

Question - My friend and I are having an ongoing discussion about the use of hive stands. He chooses not to use them. Some of the advantages I see in using them are: the hives are up higher (about 18 inches with the style I built) so there is less strain on the back when working them; it's easier to see inside them; there's less wood rot; and there's less of a problem with ants and mice.

Some of the disadvantages that my friend points out are: in cooler weather there is no natural insulation from the ground; when the bees fly in from the fields they are somewhat confused by the height and land on the stand and then have to work their way up into the hive, causing more strain on the bees than necessary.

I am fairly new in beekeeping and have much to learn, whereas my friend has been in the industry for a number of years. I respect his experienced opinion, but would still like your views on this issue. Thank you for your help.

Mark Gosswiller
Boise, Idaho

Answer - Thank you for your question. First, I'm glad you are fairly new to beekeeping because it allows you to ask any questions you want without being bogged down by "tradition".

There are pluses and minuses in both your and your friend's approach and reasoning as it relates to your job as beekeeper and the honey bee's job as honey bee.

The reason you noted as to why you built 18 inch hive stands are perfectly valid. This height will work well as long as not too many supers have to be added because you will have to reach upwards for a full super or balance on a ladder to get them on or off. That would be a good problem to have, I think.

Your friend mentions that in cooler weather there is no natural insulation, such as the ground, to protect them. Well, yes and no. Remember the bees do not heat the whole hive, just the cluster. If it is 0ºF outside it's 0ºF inside until you get close to the cluster. But if the wind is blowing, he may have a point. If not, the colony will be in contact with the cold ground which will act like a heat sink. I would disagree that bees become confused at the elevated height of the colony. Honey bees are superbly endowed with the ability to locate and see their colony. Remember also that honey bees will always choose a nest site height up in a tree or wall or rock face, etc., assuming all other things are equal. They are one of the few creatures which stores large quantities of food which other creatures want also. One of the protective mechanisms that evolved was nest site choice high above ground level. We wouldn't have honey bees if they had chose to nest on or at ground level and also stored large quantities of honey and pollen. Too many honey badgers, skunks, tigers, humans and bears would have put a stop to that.

Bees are very forgiving. If we do certain things, they will often adapt. Even if we don't do our job, they will still try to store a surplus of honey and very likely will do so.

Question - Could you explain what the speed and time should be on a Root Tangentail extractor to extract honey by hand? In a Nov. '91 article in another magazine, they state 20 minutes. Sounds wrong to me.

Joseph Reed
Mount Pleasant, PA

Answer - Your first question dealt with a hand operated tangential extractor. Remember that the tangential extractor is designed to only remove honey from one side of a comb at a time. The frame then has to be turned so that the opposite side can be cleared of honey. This method is much slower than a radial extractor which will remove honey from both sides of a frame at once. How quickly honey is removed from the comb by the centrifugal force created by the spinning extractor is also dependent on the temperature of the honey and the speed at which the extractor spins. To obtain perfectly dry combs 20 min. on a hand operated extractor with cool honey probably is true. However, generally much less time per side is required – 3-5 minutes is a good estimate.

HONEY BOTTLER

CHEAP EXTRACTING

Question - How can I extract honey without renting or buying expensive equipment?

Jethro Moine
Nags, NC

Answer - One way is not to rent the extractor, but to have someone who already has the equipment do your extracting for you on a per pound basis.

Extractors are not as expensive as they once were for the hobbyist beekeeper.

BOTTOM BOARDS

Question - The bottom board of my hive has two sides to set the hive upon. One side raises the hive up higher from the board than the other side does. If this is true do I have to turn all the boards over until the hive is resting on the shallow side?

Answer - We use the deep side of the bottom board the year around. We do use an entrance cleat in the opening for the winter. You, may if you wish, reverse the bottom board for a winter entrance.

BRANDING BEEKEEPING EQUIPMENT

Question - Recently, I have heard a great deal of conversation about branding beekeeping equipment. Why brand equipment and what units are available for branding?

Answer - Many people have had their colonies "Bee Rustled" in the past few years. Now that honey is at a profitable level, rustling is becoming more common. Most people agree that branding of bodies, supers and frames helps deter theft of equipment because it can easily be identified. The problem is so severe in some areas that non-branded frames have been stolen leaving only empty branded supers and bodies.

Brands are available from the economy brands with three 3/8" letters to a propane brand with up to five 1" letters. The small economy brands must be placed in a heat source (open fire, burner, gas flame) to brand. The propane units run off tanks or disposable bottles. In addition to these units an electric model is also available. It has up to five 1" letter capacity and is convenient since it can be used wherever there is an electric outlet. It is, however, less efficient and takes longer to build up heat between brandings.

BUILDING EQUIPMENT

Question - I plan to start on a small scale in spring more or less as a hobby. How many colonies would you recommend I start with? I am familiar with wood and metal working machines and I plan to build my own equipment. Where can I get the necessary drawings for the complete hive, extractor, etc.

Answer - We would suggest that you start your beekeeping with two colonies of bees and that you purchase two-pound packages to be installed in early April. You can go on from there depending upon how you enjoy this hobby, but we're certain you will find it fascinating.

If you really wish to build your own equipment, buy or borrow a complete outfit and make exact copies. The problem with beekeeping equipment is that the dimensions must be accurate or you will have trouble as long as you use the equipment. If the space is too large, the bees fill it with comb; if the space is too small, they glue everything tight with propolis. We seldom recommend that beekeepers build their own equipment since the savings is not great and the chance for error is fairly high. In beekeeping, you will find your time better used if you buy the equipment and spend your time working with the bees.

REDWOOD AS MATERIAL FOR HIVES

Question - During the past few months, I have gathered a lot of scraps of redwood, from which I made a number of bottom boards for my beehives. Since then I have been advised by a friend who works for a redwood distributor that redwood is bad for apiary supplies. He based his opinion on the fact that no insects live in the redwood forests and that he had also been told what he told me. What is your opinion? Also would it help if I painted the inside of the bottom boards as well as the outside?

Answer - We are not an authority on redwood nor what qualities it has that might keep insects out of redwood forests but we suspect that your friend is either misinformed or is overstating the case. We know that redwood has been used fairly extensively in the beekeeping industry as wood for the cover and also for the bottom board of beehives. It is a wood which is resistant to rot and deterioration and also is resistant to insects and the various ways in which they damage wood. However, we have never heard of an instance where redwood was harmful to bees, even when used on the bottom board. Consequently, we do not think you need to paint the inside of the bottom board although this would further protect the wood from deterioration.

DIVISION BOARD FEEDERS

Question - I purchased 10 of the Plastic Division Board feeders and am having problems. Floats were not included. I cut floats out of 3/4 inch pine. The first ones I cut went about half way down and then became stuck on the edges of the four little embossed lines that are built into the feeder apparently for structural strength. The bees then proceeded to go around the end of the float and many of them were drowned. I then cut a float smaller in width so that as the bees took the syrup it would go beyond the embossed lines and go completely to the bottom. This wasn't good either as the float was too wide and again many bees drowned. These plastic division board feeders are tapered both in width and length. In addition the plastic is so smooth that a bee floating in the syrup cannot climb up the side.

Answer - Your problem with floats in the tapered plastic division board feeders is not unique. Others have had similar problems. The taper, of course, was put in there to allow the feeders to nest for shipping, to keep freight charges down. However, they do make it difficult for float purposes. One thing you might try is using either styrofoam chips, or perhaps some of the perlite that would be available through nursery or garden centers. This is the small styrofoam beads that are used to loosen up heavy soil.

QUEEN INTRODUCTION WITH A DOUBLE SCREEN

Question - I am not successful in introducing queens. In an earlier answer you suggested the use of a double screen. I have already made three single wire screens. I know the double screens are on the market but why double? And if so, what space should the screens be apart? Would 3/8 inch be enough?

I have a hive which is weak and would like to build it up as quickly as possible and at the same time introduce a new queen. I use two deep hive bodies the year around with shallow supers above for surplus honey. I suggest the following: Take a deep hive body and a frame or two of capped brood from a strong hive and two frames of honey, and fill up the rest of the body with drawn combs.

I would take off the cover of the weak hive which now has two shallow supers on and put on the screen with the entrance at the back. Put a wire screen over the entrance for a day or two. Then take the candy out of the shipping cage and put it in the hive, including the five or six bees that come with the queen. This would be a help to the queen until the brood hatched out. If all went well, I suggest leaving the screen on for at least a month and in this time the queen would get the smell of the hive below and both queens would be working in the meantime. I would appreciate your advice on this matter.

Answer - We would think 1/4 or 3/8 inch space between the screens would be sufficient but generally about 1/4 inch is used. This prevents some communication between bees in the lower and upper hive while still making use of the ventilation or heat from the bottom hive.

Your plan should work except we probably would use about four frames of brood – two with young brood and eggs and two of sealed brood or better yet, just emerging brood. When making up divides in the same yard, where older bees will return to the original colony site, it is better to make up the new unit on top. We always release the five or six attendant bees which come with the queen as it seems to increase the chances for success in requeening. This should all be done over a double screen so you have the combined efforts of the two queens. Remove the screen after a month.

EXTRACTOR

Question - I have five colonies of bees, but do not get enough honey to use an extractor. Last year I put the honey and comb in a double boiler, heating and crushing the comb while stirring, not allowing the honey to get over 150º F., but the honey never comes out clear.

Would you be kind enough to help me get clearer honey? I was wondering if the honey comb could be crushed in cheesecloth, without heating and let drain at room temperature.

Answer - You would get a better quality honey if you crushed the comb and allowed it to drain through cheesecloth in a warm room. We don't believe that you can afford to get along without a small extractor. It wouldn't take many sessions of melting or crushing combs to pay for the extractor, if only the cost of the foundation were taken into consideration. Save yourself a lot of mess and get larger amounts of better quality honey by obtaining an extractor.

FOUNDATION

Question - What is the best way to nail foundation into the frames? Should the nails go in from the side or from the bottom to the top?

Answer - Put the nails in the top bar at a slight angle from the bottom to the top. This holds the strip of wood tightly against the foundation and frame.

ONE OR TWO HIVE BODIES

Question - Should each stand of bees have more than one hive body? Wouldn't more than two make it difficult to find the old queen?

Answer - Most colonies of bees do better if they have two hive bodies for the brood chamber. For honey storage, either additional hive bodies or shallow supers are used.

It is true that the more units there are in a hive, the more difficult it is to find the queen. Many beekeepers use a queen excluder between the brood chambers and the supers to keep the queen in the brood chamber only.

INNER COVERS AND BEE ESCAPES

Question - I purchased a single hive of Midnight package bees last spring essentially to help the pollination of the several apple, peach, and pear trees and grape, barberry, almond, and filbert shrubs which I have on my lot. Should the hole in the inner cover be covered up in the winter? And which is the top and bottom of the inner cover?

Answer - The hole in the middle of the inner cover is put there for ventilation purposes as well as to allow the inner cover to double as an escape board. A bee escape, a small device that works like a gate, allowing only one way traffic through it, is insert-

ed in the hole in the inner cover. The inner cover then is placed beneath the honey supers to allow the bees to escape down to the brood chamber. As the bees cannot return through the bee escape, the honey super will be comparatively free of bees by the next morning.

Many beekeepers also use some kind of a channel device to put over the hole in the inner cover leading to the outside so they have a "guarded" upper entrance for winter. When we use the inner cover here, we normally use the shallow edge of the inner cover down next to the frames. This allows less space for the bees to build burr comb if they should get crowded. Also, the deeper rim on top allows us to feed dry sugar to the bees on the top of the inner cover if we choose to do so.

PAINTING

Question - I am just getting started in beekeeping and have a question. I am thinking of painting the inside of the hive bodies, and am wondering if there would be any adverse effect on the bee colonies from doing this.

The purpose of the paint would be to prevent warping. Since there are many types of paint available, are there any that should be avoided?

Answer - Most references advise against the painting of the inside of the hive bodies. One of the reasons given is that the moisture which accumulates within the hive, is more apt to condense on the walls of the hive and continue to be too humid. The unpainted sides of the hive bodies can absorb some of the moisture and thus make the inside somewhat dryer.

QUEEN EXCLUDER

Question - I have a hive of bees with a double brood chamber. On top of the second chamber, I have put a queen excluder. After the excluder, there are three shallow supers. Should this excluder be kept in the hive during the winter? If I remove it, will the queen (she's new) lay in the shallow supers? If she should get into the top shallow supers, what could I do about it?

Answer - Remove the queen excluder in the fall. The real danger of leaving the excluder on the hive all winter is that should the cluster move through the excluder in search of food, the queen would have to stay below the excluder where she would die from lack of food and the cold. The queen has tapered off considerably in her brood rearing at this time, and would not be so apt to go up into the shallow supers. This would be particularly true if the fall flow were very intense, and the bees were storing nectar up there.

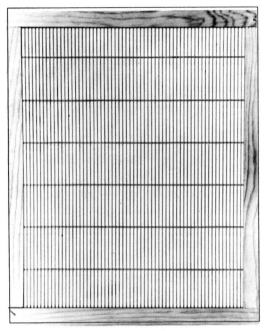

QUEEN EXCLUDER

RECONDITIONING USED EQUIPMENT

Question - We recently purchased a quantity of used bee equipment including hive bodies, supers, bottoms, lids and some very good frames. We would like to know how we can be sure they are clean. Can we boil the frames and excluders in lye water? We have scraped the supers and bodies clean and used a blowtorch on them. Is that enough?

Answer - You have correctly handled the used bodies and supers which you purchased by scraping them clean and using a blowtorch inside. The frames can be boiled as you have mentioned to sterilize them using about one can of lye to 5 gallons of boiling water.

SHALLOW SUPERS FOR EXTRACTING

Question - Why is it that small, as well as big honey producers, have gone into shallow supers for extracting? Shallow supers are almost unknown here in Argentina and are supposed to bring only double handling with no advantages. Colonies are worked here in 3 full depth bodies as a rule. Honeyflows are generally slow and very long, but in good years there are 3 separated "rush" periods in a season in which bees can fill a full depth super in a week.

Answer - *There are several reasons for using shallow supers for surplus honey, but here are the two reasons heard most often: 1) the supers are filled more completely and sealed more quickly, and 2) they are lighter in weight to handle. These are good reasons, especially if you have light honeyflows and like to control the space more closely for storage. It takes a strong back to handle full-depth supers all day.*

In Argentina and in certain parts of the United States, it would be much better and less work to use full-depth supers, since you would have fewer supers and frames to handle. A large number of our commercial beekeepers use the full-depth supers exclusively.

SOLAR WAX EXTRACTOR

Question - Will a solar wax melter be worth the investment?

Answer - *During the hot months of the year a solar wax melter can help you save the wax that is in all those scrapings from the hives, the pieces of burr comb the bees manage to find places for if they get a little crowded, and the broken and poor combs that you discard. So often there is such a small bit of material at any one time that it is neglected or burned up just to get rid of it. Yet these same little bits, when thrown into a solar melter setting at the edge of the beeyard or beside the honey house, can render a surprising total amount of wax.*

There is very little work involved. Just lift the glass top and throw in the material to be melted. The sun does the work for you. When the wax pan gets full enough, empty it when the wax is congealed. Occasionally the slum (the nonwax residue) must be scraped out too.

DADANT STAINLESS STEEL SMOKERS

USE OF THE SMOKER

Question - I'm having real trouble using the smoker and examining the hive. I start out pretty well and then usually the smoker goes out, the bees get excited, and my orderly plan for examining the hive falls apart.

Answer - The smoker is a very essential tool, one which many beginners use improperly. Many materials are satisfactory for fuel, as long as they are easily available, easy to light, porous enough to allow sufficient draft so the fire doesn't go out every time you set the smoker down, and as long as the burning fuel does not smell so horrible that neither you nor the bees can stay in the beeyard. We use mostly old burlap bags even though they don't smell as nice as pine needles.

You must keep enough fuel in the smoker so it is really a smudge pot, not a fire pot – so that the smoke is cool and not hot. Hot smoke usually accomplishes just the opposite of what you want: instead of calming the bees by causing them to stick their heads in cells of honey to fill their honeysacs, the hot smoke causes them to become angry. Excessive and unnecessary pumping of the bellows creates a hotter fire, and in turn, a hotter smoke. This also helps to shorten the life of the smoker.

When you have finished using the smoker, the best thing to do is to empty the fuel and coals onto the ground and pour water on them until you are certain no fire is left. In case this is not advisable, both the smoke nozzle and the air intake hole at the bottom should be stuffed with green grass or leaves to suffocate the fire. A couple of corks to fit these holes are good for this purpose. One reminder – a smoker which is not emptied or properly stuffed can become a real fire hazard.

How much smoke should be used? There is no one answer. Generally, less smoke is needed on warm days, probably because propolis becomes softer and there is less popping and cracking when opening the hive, and because the bees are more apt to be collecting pollen, water, or nectar.

When opening the hive in cool weather, a few puffs of smoke directly into the entrance first thing gives you the jump of the guard bees. Then if the lid and inner cover are stuck with propolis, use your hive tool to pry as gently as you can. As soon as you have just a little opening, puff in a little smoke. Pry a little, puff a little – an occasional puff or two at the entrance. These recurring smoke puffs are like a boxer with a good left jab, continually pecking away, keeping his opponent from getting set.

With the inner cover off, the next step is usually to take out a side comb with a few or no bees on it. This allows the other combs to be pried sideways as they were loosened with the hive tool. Then, as they are lifted up, the bees on the two adjoining combs are not "rolled" or mashed together, thus minimizing the disturbance (and the chance of injuring the queen). As you become better acquainted with how bees react to intruders and to smoke, you will develop a "sensing" of their mood. From this, you anticipate a need so that you "jab" them in the face with some smoke just before they need it. Until this has become well learned, it is better to puff the smoke a little too often, rather than not enough.

Then, each individual frame with honey in it can be lifted out, bees and all, and then the bees can be brushed off in front of the colony using a bee brush. Now it would be nice if we could guarantee you that these methods will always work and that you will be able to examine your bees under any conditions without overly disturbing them but such proper use of your smoker will go a long way in the right direction. At first wear a veil and bee gloves and tie a string or rubber band around trouser legs or tuck the legs inside your socks. It is also advisable to work with the bees in a rather slow manner with even movements. Fast, jerky movements will have more of a tendency to aggravate the bees. Likewise, always work the colony from the side as standing in front of the entrance, or approaching a colony from the front, make you a direct target.

SMOKER FUEL

Question - What is the best smoker fuel you have encountered?

Answer - Nearly every fuel has been tried from red peppers to tobacco. There is always something new for smoker fuel but many beekeepers return to using overalls or blue denim. It doesn't smell unpleasant, calms the bees, is long burning, and produces a great cloud of smoke. Should there be any doubt just try a roll about the length of the smoker tank and about three or four inches in diameter. Insert the lighted end of the roll of denim in the smoker, keep the fire going with an occasional puff on the bellows, and you will find pure cotton denim most satisfactory.

CLEANING A SMOKER

Question - I have used the same smoker for two years. It is very black and sticky inside and it is very hard to get anything to burn in it. What is the best way to clean the smoker and is there anyway to prevent the smoker from becoming black again?

Answer - The type of fuel you use will determine the amount of black deposit in your smoker. Fuels such as corn cobs, rotten apple wood and similar woods low in pitch content, will burn without creating the problem. The best way to clean out the black deposit is to scrape it out as much as possible and then wipe the insides down with a rough textured cloth. The grate should be removed and scraped. The air holes in the grate should be opened up and a rattail file is good for this. If you repeat the treatment as needed, you should be able to keep your smoker going for years.

SUPERS

Question - I'm just starting beekeeping, I have some old supers. A man told me I have to wash them in salt water. Another told me I have to char them all. Which should I do?

Answer - *The only reason for making an attempt to clean up those old supers in the first place is if you suspect that they may be contaminated with the disease organisms from the foulbroods. Washing in salt water would very likely have no effect, although the suggestion to char the inside of the super shells would be effective.*

To be doubly safe, preventive treatment of doses of Terramycin might also be suggested, as long as feeding of the drugs was discontinued thirty days before the start of the honeyflow.

BEEKEEPERS COVERALLS

WHITE CLOTHING FOR BEEKEEPERS

Question - Do beekeepers wear white because bees recognize this color?

Answer - *It is true that beekeepers wear white clothing to enable them to work their colonies of honey bees without being stung to any extent. This is because the compound eye of the honey bee recognizes the movement of a dark object more quickly than one which is light colored or white. This movement of a dark object excites the bees to protect themselves and this often results in stinging. For example, a beekeeper working bees will often remove a wrist watch because angry bees are inclined to head toward it. In a similar way, they will fly toward the eyes or the dark opening of a mouth in much a similar way. This is why beekeepers wear bee veils. Similarly a* *beekeeper wearing white socks will not be stung as often around the ankles as would a beekeeper wearing black socks.*

The vision of the honey bee is quite different from that of man. The colors red and yellow at that end of the spectrum are seen as grays and even black objects and, therefore, the bee is color blind with respect to these colors. The bee does see the greens, the blues and the violet colors and then continues to see in the ultraviolet range where man does not see. Colin G. Butler, in his book entitled "The World of the Honeybee" states, "Some white objects which appear equally white to us may appear as of two quite distinct colors to bees – the first can readily be distinguished by them probably as a shade of blue-green, whereas the other kind of white object probably appears as a shade of white." He continues, "Therefore, since most white flowers absorb ultraviolet light, such flowers probably appear to be blue-green to bees."

We do not know whether the white cloth of a uniform for beekeepers would reflect ultra-violet light or not but have been informed that ultra-violet light is used to determine the whiteness of clothing in studies of various detergents used for cleaning them. It is claimed that the use of ultra-violet light enables one to tell whiteness better than when ordinary light is used. Who knows, perhaps the white uniform of a beekeeper might also appear blue-green to bees.

**BEESWAX AS IT IS SENT TO DADANT & SONS,
BY BEEKEEPERS**

BEESWAX SHOE POLISH

Question - I read your article on furniture polish with great interest in the December issue of ABJ.

I have been looking for some time for a shoe polish made from beeswax. Also, a leather waterproofing for leather boots and shoes made from beeswax.

If you have such a formula, I would appreciate receiving it.

R.G. Burggraf
Waverly, Ohio

Answer - I had to do some digging for this question. I finally found what I think you are looking for in an old book about beeswax from England. Because of the age of the attached recipes, some of the ingredients may be as obscure to you as they were to me. I've gotten you started at least. Best of luck. Let me know how it turns out.

Wax Shoe Polish - Melt together 1 lb. of white wax, 1 lb. crown soap, 5 oz. ivory black, 1 oz. indigo, and 1/2 pint nut-oil. Dissolve over a slow fire, stir until cool, and turn into small moulds.

Waterproofing Boots and Shoes - Beeswax, 1 oz.; suet, 1/2 oz.; olive-oil, 2 oz.;

lampblack, 1/2 Melt the wax and suet in the oil, add the lampblack, and stir till cool; warm the shoes and rub in the compound.

Waterproof Harness Paste - *Put into a glazed pipkin 2 oz. of black resin, place on a gentle fire till melted; add 3 oz. of beeswax, and when this also is melted remove the mixture from the fire; add 1/2 oz. fine lampblack and 1/2 dr. of Prussian blue in fine powder. Stir till mixed well and add sufficient spirits of turpentine to form a thin paste. When cool apply a thin coat of the paste with a rag and polish with a soft polishing brush.*

Polish for Harness - *Melt 8 oz. of beeswax in an earthen pipkin and stir in 2 oz. ivory black, 1 oz. Prussian blue ground in oil, 1 oz. oil of turpentine, and 1/2 oz. copal varnish. Make into balls. Apply with a brush and polish with an old handkerchief.*

TEST YOUR BEESWAX KNOWLEDGE

True or False

1) Bees consume about 8.4 pounds of honey to produce 1 pound of wax.

2) World demand for beeswax far exceeds the supply.

3) Most wax is produced by house bees that are two weeks of age.

4) A "No Pest Strip" is a good way to protect comb from wax moths.

5) Dusty appearance that forms on beeswax is a form of mold that is easily wiped off.

6) The melting point of beeswax is approximately 125 degrees F.

7) Bees produce beeswax from a pair of wax glands on the ventral side of their abdomen.

8) The color of beeswax is white when secreted by the bees. The yellow color comes from the contact with pollen.

9) The advantage of beeswax candles is that they burn slowly without smoking, drip less, have a bright light, and have a mild, sweet scent.

ANSWERS TO WAX TEST

1) True

2) True

3) True

4) False - Wax will absorb chemicals, such as pesticides, and should not be stored near them.

5) False - Dusty appearance is called "bloom". It is easily wiped off. It is the waxes of a lower melting point migrating to the surface. Bloom melts at 102 degrees F.

6) False - Approx. 145 (Actual 147.9 + or - 1 degree)

7) False - There are four pair of wax glands

8) True (Actually clear)

9) True

BEESWAX FURNITURE POLISH

Question - Could you tell me how to make a good furniture polish?

Bill Butt
Butte, Montana

Answer - This is a popular question this month. Several letters have arrived asking for the same information. I have no personal experience with any of these polishes, so my judgement of their merit will be left to you.

Furniture Polish - 1 pint linseed oil, 2 pints turpentine, 1 to 2 ounces of beeswax. Combine oil and beeswax, remove from heat and add turpentine and mix. Shake well before applying.

Furniture Polish - 1 gallon soft water, 4 ounces soap, 1 pound beeswax, 2 ounces potassium carbonate. Dilute with water to suit preference.

Furniture Polish - (Paste form - best for old furniture.) Use 1/2 pound turpentine and 1/2 pound beeswax. This also makes a satisfactory floor wax by varying the amount of turpentine added.

Floor Wax - 1/2 lb. beeswax, 1/2 pint turpentine, 1/2 cup alcohol. Melt wax over hot bath. When melted stir in turpentine and alcohol. Stir until mixture is a thick paste and pour in jars.

Floor Wax - 1/4 lb. beeswax, 1/4 lb. paraffin, 1/2 pint turpentine, 1/4 pint alcohol. Melt wax and paraffin over hot water bath. When melted stir in turpentine and alcohol. Stir until mixture is a thick paste and pour in jars.

Floor Polish - 1 pint turpentine, 4 oz. beeswax, 3 oz. ammonia water (10% strength), 1 pt. or less water. Melt turpentine and beeswax over hot bath. When melted remove from heat and stir in ammonia and water until cream is cool.

Question - What proportion of well drained honey cappings will result in good beeswax? Shouldn't it be almost 100%? The person who renders my cappings now is only getting about 70% beeswax in comparison to the original weight.

Mark Meister
Redmond, WA

Answer - If the beeswax cappings were <u>perfectly dry</u> with no honey sticking to them at all and no other "stuff" mixed in with the cappings, then you would think that you should get, pound per pound, the same amount of wax back as you brought in . The key phrase here is "well drained". What does well drained mean? To me it means having your cappings in a container with a screen or sieve bottom which allows the honey that was caught in the cappings during the uncapping process to drain down and out by gravity. Honey draining by gravity is affected by not only the force of gravity but by the temperature; warm honey flows faster than cold honey, and by the density or how tightly packed the cappings are. Gravity is not strong enough to completely strip off all the honey that is adhering to the cappings, especially cold honey. Honey can be trapped in little pockets all through the cappings. Honey weighs about

12 lbs. to the gallon. To have a gallon of honey trapped even in a small amount of cappings is not unusual. All this weight can throw your calculations of wax available off quite a bit. In addition, you must count all other foreign material - pollen, propolis, dead bees, grass, etc.

I think getting 70% of the original weight you brought in is quite good. We often hear of rendered wax returns of a much smaller percentage.

USE CAUTION WITH BEESWAX!

When making formulas containing turpentine, alcohol, oil or wax, remove away from heat when adding ingredients.

Marian Chandler
W. Caldwell, NJ

Question - In the past you have reprinted several recipes for beeswax furniture polishes and floor waxes. Can you suggest anything for shoe and boot waterproofing?

I see some of the commercial product available are composed primarily of beeswax. I feel silly buying it when I have several hundred pounds of wax in storage.

Oliver Frank
San Mateo, CA

Answer - Here is a recipe I found in a very old English book on beeswax. Some of the ingredients may seem peculiar in our world of synthetics. I liked it because it certainly is all natural and environmentally friendly.

If you have a dog, I'd keep your shoes and boots away from him or her because the ingredients may turn your footwear into snack food for Fido.

Wax Dubbing - Melt together 6 1/2 parts yellow beeswax, 26 1/2 parts mutton fat, 6 1/2 parts thick turpentine, 6 1/2 parts olive oil, 13 parts lard, stirring into this 5 parts of well-heated lampblack. The mass is then poured into little tin boxes for convenience in using. The dubbing is first warmed and then applied by being rubbed in well with the fingers. Hard leather is thus softened and becomes perfectly waterproof.

BURR COMB

Question - I am a beginner and have 5 hives of bees. I have them in two 10-frame deep supers for brood nests. When I manipulated them, I found quite a little burr comb on the frames. Should this be scraped off? If so, when should it be done?

Answer - Burr comb should be scraped off whenever you find it as it is only going to be a nuisance. The hive tool is excellent for removing burr comb and if you remove what burr comb you see every time you examine the hive, yours will be kept relatively free and much easier to manipulate. Save the wax. If you throw it on the ground, it may be the cause of robbing. A solar wax melter is an ideal unit for melting the bits and pieces of beeswax. It gets them out of your way and you'll be surprised at the amount of wax that will accumulate.

SOLAR WAX MELTER

Question - I am an avid reader of the ABJ. I have several questions concerning beeswax. I hope you can briefly answer these questions.
1) Why do bees produce beeswax?
2) How do bees produce beeswax?
3) What is the composition of beeswax?
4) What is the melting point of beeswax?
5) Why are there different colors of beeswax?
6) Is beeswax produced in the U.S. different than beeswax produced in other parts of the world?

Dave Coovert
Union, MO

Answer - I don't think I can fully answer any of your questions briefly, but I'll try.
1) In response to large amounts of available nectar or sugar syrup provided by the beekeeper, bees about 14 days old start producing beeswax for comb building.
2) There are four pairs of wax-secreting glands on the ventral (bottom) side of worker abdomens.
3) Beeswax is a complex mixture of lipids (fats) and hydrocarbons. With the advent of gas chromatographic equipment, over 300 individual chemical components have been identified from pure beeswax.
4) Beeswax melts at a range of 62-65º C. or 144-149º F.
5) The different kinds of and percentages of sugars in the nectar or sugar syrup they feed on influence the color of the beeswax produced. Colors range from almost translucent to lemon yellow in freshly produced beeswax. Color is also affected by flower or tree sources being foraged upon as well as how much filtering the beekeep-

er or beeswax processor does. Heat also affects wax color.

6) Beeswax produced from Apis mellifera is generally similar throughout the world. Remember that other "bees" produce "beeswax" which is chemically different from that produced by honey bee. These "bees" include bumble bees, Apis cerana, Apis florea, Apis dorsata and several of the Central and South American stingless bees.

**WORKER WITH NEWLY FORMED
BEESWAX FLAKES ON ABDOMEN**

BEESWAX FLAKES FROM HONEYBEES

COMMENT

I'm continually amused at the interest in beeswax-based polish formulas based on the number of inquiries and suggestions I get on the topic. As I get good information, I'll pass it on to our readers.

This information comes from Elaine White of Starkville, MS.

Beeswax polish with its soft, satin shine once was considered the ultimate in wood care. But, it is fast losing this distinction due to poor products being sold and good products being sold without proper instructions for their use. There is also confusion between beeswax polish formulas designed for bare wood surface and formulas designed for sealed wood surfaces.

Formulas consisting of beeswax, turpentine and linseed oil are designed to provide a hand-waxed finish to bare wood. This old-fashioned and labor intensive method of finishing wood involves the application of multiple layers of wax and friction polishing between each application.

Modern woodwork is almost always sealed with varnish, shellac, paint or synthetic finishes. Most polish formulas for sealed-wood surfaces contain beeswax and turpentine; the amount of turpentine or mineral spirits determining if the product is a liquid or a paste.

FORMULAS

Liquid Polish I - 1/2 cup beeswax (about 4 oz. wt.) Melt the waxes on high in a microwave (watch closely) or in a double broiler. Remove from heat and stir in the mineral spirits.

Paste Polish I - Use the Liquid Polish I formula reducing the amount of mineral spirits to 1 1/2 cups.

Liquid Polish II - 1/2 cup beeswax (about 4 oz. wt.), 2 tablespoons carnauba wax, 1/8 teaspoon lye, 2 cups water, 2 cups mineral spirits. Melt the waxes on high in a microwave (watch closely) or in a double boiler. Add lye to the water and stir until it is dissolved. Remove the wax from heat and add the lye water. Immediately add the mineral spirits while stirring briskly or by using an electric mixer.

Paste Polish II - Use the Liquid Polish II formula reducing the amount of water to 1/4 cup and mineral spirits to 1/2 cup.

Cream Polish - Make Paste Polish II and allow it to cool. Make small additions of mineral spirits while whipping with an electric mixer or blender until the polish has the consistency of hand cream.

WARNING - Work outside or open windows and doors for adequate ventilation when mixing mineral spirits. Due to danger of fire when heating wax and mineral spirits, do not use direct heat.

Question - Does beeswax have any health benefits? I eat it with comb honey on bread and have never felt or read that there are benefits.

Fred Fulton
Montgomery, Alabama

Answer - Beeswax is not digestible. I suppose because it is indigestible, it acts in a similar fashion as roughage which we are told to eat more of these days. Beeswax does have small amounts of pollen in it which would be digestible.

CHAPTER 8

HONEY PLANTS

Question - I have a 2 acre field near one of my honey-bee colonies that I have decided to plant with white sweet clover seed (about 12 lbs.) for the bees to utilize for some high quality clover honey. What is the best method of planting this seed? I have read that you can sow the seed on top of the snow in very early spring and it will plant itself when the snow melts. Do you have any suggestions for me?

Tim Armstrong
Ashtabula, OH

Answer - I went through a similar situation last year only I did about an acre. I planted a couple of varieties of clover and birdsfoot trefoil.

Yes, you can seed on top of the snow or in early spring right on top of the ground. I would suggest that you also purchase the correct inoculate (the fungi that attaches itself to legumes). Mix it well with the seed and try to evenly broadcast it. I use a hand-held device where you crank the handle and a spinning disk picks up the seed from a hopper and throws it out. I'm guessing, but it seemed that I had a germination rate of around 70-80% using this method.

WHITE SWEET CLOVER

POISONOUS HONEY PLANTS

Question - At one time I had heard that Azaleas make honey poison. Can this be confirmed? And then Rhododendron being a broad leaved evergreen growing in similar conditions as Azaleas, do they have the same problem?

Richard Shoots
Westerville, OH

Answer - I went to our publication American Honey Plants and found this: "The Azaleas are closely related to the mountain laurel and are likewise sometimes report-ed as poisonous. The flame-colored Azalea is common in the mountains of the east-ern states from Pennsylvania to Georgia. It has a profusion of showy, flame-colored blossoms coming just when the leaves appear. Although there are several species of Azalea or Rhododendron coming from eastern Canada to Georgia, there are few reports available which indicate that they are anywhere regarded as important nectar plants by the beekeepers."

What I gather from the above is yes poisonous honey can originate from Azalea (Rhododendron), but there are few places where the concentration of flower-ing Azaleas would lead to a surplus of pure Azalea honey.

Question - If I tap maple trees and get the sap, can I feed it to the bees instead of sugar water?

Will the bees take it or will it hurt the bees in any way?

Charles Damron
Wyandotte, MI

Answer - If nothing else is available, the bees will take it, but we don't suggest you feed it. The sugar concentration is not great as anyone making maple syrup knows. I think it takes 40 gallons of maple sap to make 1 gallon of syrup valued at $20.00. At $20.00 or so a gallon, I'd certainly make maple syrup and buy cheap sugar to feed the bees. It also may cause dysentery similar to that caused by honeydew stored by bees for over-wintering. The high dextrin content of these sugar sources gives the bees dysentery because their bodies cannot digest these materials.

Question - I am trying to locate for purchase a shrub called: Gallberry (Ilex glabra). I have inquired with my local agricultural extension agent and several nurseries, but they were not familiar with this species. I understand, through my reading, that this shrub is a heavy honey producer and should be hardy to the foothills of western North Carolina.

I would greatly appreciate any information that you might be able to share with me regarding this matter.

Dwight Tew
Franklin, TN

Answer - Dr. George Ayers of Michigan State University was most helpful in identify-ing where to look for a source of the plant Gallberry (Ilex glabra). There is a little known reference work called "The Andersen Horticultural Library's Source List of Plants and Seeds." What this book does is list the plants in one section with numeri-cal coding that sends you to another part of the book listing suppliers. I didn't know anything like this existed before I asked George for help on your question. It is a real treasure. Here are some suppliers listed in "Andersen's" for you.

Woodlander's Inc.
1128 Collection Ave.
Aiken, SC 29201
803-848-7522

Shady Oaks Nursery
700 19th Ave. N.E.
Wasela, MN 56093
507-835-5033

Girard Nurseries
P.O. Box 426
Geneva, OH 44041
216-466-3705

Gorest Farm
990 Tetherow Rd.
Williams, OR 97544
503-846-6963

Sunlight Gardens
Rt. 1, Box 600-A
Andersonville, TN 37705
615-494-8237

These should give you a start in finding your honey plant.

Question - I planted Basswood and Black Locust as an additional source of nectar for my bees. When in bloom the trees swarm with seemingly every imaginable insect except the honey bee. WHY?

R. D. Zacios
Ont., Canada

Answer - Here are several possible answers that individually or in combination may be correct.

1) Nectar production in most all bee forage is not understood very well at all. Sometimes nectar/pollen is produced, sometimes it isn't. There is no general consistency to rely on. Perhaps there is not enough nectar or pollen to offer honey bees, so there is no reason for them to be there.

2) Because of the devastation of honey-bee feral colonies as a direct or indirect result of mites, there may not be a large population of honey bees to allow visits to your trees. Perhaps there is something better to collect nectar and pollen from at this time for the bees.

3) Obviously, there must be something in the flowers to attract the other insects you mention. Perhaps these insects are collecting the nectar/pollen resources and drawing them down before the honeybees can take advantage of them. As a result, the bees are going elsewhere.

I'm sorry that there is not one nice clear definitive answer for this problem. The only consolation is that you are not the only one to notice it.

Question - I have hives on 480 acres of a mixture of white ladino clover, white Dutch, arrow leaf and red clover. By watching the bees - they never seem to touch the red clover. Can you explain this?

George Richtmeyer
Noble, OK

Answer - Have you ever dropped something and try as you might to reach and squeeze and press for it, the item is still just a few inches out of your reach? No matter what you do or how you position your arm or your body, the item you want is still just inches out of your reach. After a while you have to face the fact that nothing you do with your body is going to allow you to reach this item. You give up in the quest and expend your energy in other pursuits.

If you look at a white ladino clover flower head and a red clover flower head, you see immediately that the red is larger than the white. It is larger because each floret on the red clover head is larger and most importantly longer than the individual white clover floret. The honey bee's tongue is just not long enough to reach the nectar at the bottom of the red clover floret corolla tube. No matter how hard the honey bee pushes and extends its tongue and its body, it isn't enough. So it gives up and faces reality that, energy expenditurewise, it isn't worthwhile to devote attention to red clover.

You've probably noticed that bumble bees frequent red clover and hardly ever white if its in bloom at the same time. Their tongue is long enough to gather nectar from red clover. They could also gather nectar from white or Ladino clover, but why compete with honey bees when they can have the red all to themselves?

WHITE DUTCH CLOVER

Question - Although I am a Master Beekeeper, I am a hobbyist with two small bee-yards in central and northern Baltimore County, Maryland. Our honey flow occurs with the flowering, in the second half of May, of black locust and tulip popular. Black locust blossoms a day or so ahead of tulip poplar and, therefore, the bees become constant to it and rarely get to the tulip poplar.

This spring the bees were storing black locust nectar in spades, judging by the weight increase, until we had a week of cold rain. While they were able to harvest the black locust nectar, I examined the nearby trees at different times of the day and observed little or no activity by the bees on the blossoms. I was embarrassed that I did not know in what part of the day the flowers produced their nectar.

I found nothing in the literature that gave me a hint. The detailed article by George and Sandy Ayers entitled "Designing a Bee Forage System - the Development of a Short List of trees (part I)", *American Bee Journal*, May, 1996, is silent on the question. Experienced beekeepers in the area did not know the answer, and, for the most part, simply shrugged. It seems to me this could vary from locality to locality and sometimes be important information for a beekeeper to have in mind. Can you help?

William J. Evans
Baltimore, MD

Answer - *You are absolutely right that information about when a plant will secrete nectar would be extremely valuable to beekeepers. It would make hive management and movement much more efficient. But, (there is always a but isn't there) there is no answer as I found out.*

I didn't know the answer either so I went to Dr. George Ayers. He told me that the understanding of the when, what and how of nectar production in honey bee forage plants is the least understood of anything in apiculture. No one has made a study of this most important question and no one probably will until beekeeping is an industry on a more solid foundation.

So, you were right in your survey. Nobody really knows.

Question - I have a few acres of land and would like to plant it this spring in something that will be beneficial for my bees and give me a better honey crop. What do you suggest?

Max Kearse
Kirksville, MO

Answer - *Rather than assuming that I know the best "honey plants" to tell you to plant, I went to the most knowledgeable honey plant person I know, Dr. George Ayers of Michigan State University.*

He told me that the surest honey plants are the legumes. Plant a mixture of alfalfa, clover and birdsfoot trefoil and then manage for bee forage. George said the reason for this mixture is that we have hundreds of years of experience in growing this stuff. We know what varieties in what part of the country adapt best to climates and soil types. We know how to plant it, keep the weeds out and harvest a hay crop. You may be able to make a deal with a local farmer to plant this for you and then allow it to bloom freely before cutting for hay, share in the profit and allow it to bloom again and harvest hay again. Where you are in Missouri, you should be able to get at least two blooms and cuttings, maybe three. Other areas of the country could get more.

George also suggests some plants that we don't know quite as much about management as the legumes. These are vitex, anise hyssop and basswood. These would be longer blooming plants and in regards to basswood, a possible lumber source sometime down the road. Planting, varieties, availability, preferable soil, weed control, etc., are kind of sketchy on these and may take some experimentation on the grower's part. I also suggest you read Dr. Ayer's chapter entitled "Bee Forage of North America and the Potential for Planting for Bees" in the 1992 edition of The Hive and the Honey Bee. It's the most up-to-date and best source of information on honey plants.

SOYBEANS AND HONEY

Question - I have been a beekeeper since a stint in the Peace Corps in th early 1970s and still keep a dozen hives or so for various therapeutic reasons. I have a small acreage and sow 20 acres of soybeans on an annual basis. What soybean variety would be best to grow in central Missouri that would yield the most nectar? With a small acreage I can usually sow the beans about the first to the fifteenth of May in this area. That is considered early around here. Sweet clover and milkweed are the meat of the surplus in this area and it would be nice to increase my bean yields and give the bees something extra to do. I walked the beans when they were flowering this year, which was after the clover flows, and didn't really see much activity. Williams is a common variety in this area. Could I grow a more conducive variety?

David R. Cole
Fayette, MO

Answer - The last significant research of which we are aware concerning soybeans and nectar production was conducted back in the early to middle 1970s by Dr. Eric Erickson. Dr. Erickson at that time found that bees visited the soybean varieties Adams, Corsoy, Hark, Illini, Lincoln, Wayne, and Williams with the most frequency in Southern Wisconsin.

In another article by Dr. Erickson, he found that soil conditions also affected honey production from soybeans. His research suggested that heavier soils having a pH of 6.0-7.0, high fertility, including potassium, and ready water availability favored maximum soybean production and substantial honey crops.

As in all aspects of agriculture, the variety grown, soil conditions and weather based partly on the geographical location you are in all have a bearing on nectar production in soybeans. There is no easy answer, but with some experimentation along with guidelines listed, you may be able to optimize your soybean honey production on your 20 acres.

Question - I would like to know if a bush we call Devil's Walking Stick which has a large flower attractive to honey bees is a good honey plant.

Marion Shelton
Rutherfordton, NC

Answer - Here is what I have been able to find out about the plant called The Devil's Walking Stick, Devil's Club, Hercules Club and Angelica Tree (Aralia spinosa):

The plant with the names above is a shrub or small tree common to damp borders of the woods and river banks from Virginia to Missouri and south to Florida and Louisiana. The flowers are white and appear in early summer. It is reported as yielding nectar abundantly, though not often as a source of surplus.

It blooms at the same time as buckwheat and yields best in seasons which are too dry for buckwheat to do well. The honey is said to be light in color, of good body and fair flavor. The plant may be readily recognized by the large number of peculiar thorns.

Question - I've just started beekeeping in Florida. I've noticed the bees seem to be very attracted to the flowers on the tree in the picture that I've enclosed. Could you identify it for me and tell me if it is a good honey tree?

Regino Arias
Hialeah, Fl

(Photo Not Available)

Answer - The picture of the flowering tree that you sent brought back memories of my childhood. As a boy in South Florida in the early 1960s, we had two of these trees in our front yard. This was before I had any idea about honey bees. These are Melaleuca Trees (Melaleuca leucadendra).

These trees were brought over from Australia over fifty years ago. They are frost sensitive, so only can grow in the warmer parts of Florida and California. They don't mind swampy areas at all and thrive there. The wood is supposed to be very close grained and will last very long underground or in contact with water. It is used in some wooden ship building and piling.

It blooms in late summer or early fall and produces a good nectar flow for 4 to 6 weeks. The honey is amber, granulates quickly and has a distinct aroma.

That's the good news for beekeepers.

The bad news is for the environment of Florida. This is a species that was introduced from Australia. It was brought to a part of our country that is warm and has plenty of water like Australia. What didn't come over with the tree was disease, parasites, and animals that eat this tree. As a result, from the time this tree was in low numbers in the 1960s when I remember it to now, it has expanded its range dramatically. It is growing profusely in Florida and, in particular, in the Everglades National Park, crowding out native plants right and left. I've seen pictures where the trees formed a solid canopy for acres and acres. They grow so quickly and thickly that they take over complete areas, replacing them to the extent that native wildlife cannot live, grow and exist in these areas.

Botanists have been trying to identify natural control measures for Melaleuca for several years without a lot of success. It would be nice to have a balance in nature and for the beekeeper. I'm sure there soon will be a workable answer.

CANOLA HONEY PRODUCTION

Question - I have 10 colonies of bees. This year I have the opportunity to put them on a canola field. I would like some information on what to expect, such as the quality and quantity of honey.

Charles Damron
Wyandotte, MI

Answer - I am glad that you have the opportunity of placing colonies of honey bees of a Canola field. Canola, being a member of the mustard family, blooms early in spring. The size of the field will, of course, partially determine the amount of nectar your bees will be able to collect. If your bees will have access to a couple hundred acres, then you can expect a couple supers of honey from each of your 10 colonies if your Michigan weather will oblige. The honey will be light in color and mild in taste, but will granulate very quickly. Keep an eye on your supers and remove and extract when they are 80% capped. If you don't you may find the honey granulated in the comb and you won't have any for yourself.

BEE-BEE TREES

Question - I planted a lot of Bee-Bee trees and the first one at home is in full bloom with clusters all over it. The bees have worked it since the 5th of July. Can you tell me if the seeds will be on the bloom clusters or where? How would you handle them for planting in the spring? I would like to start a project with the Boy Scouts and plant all of the worst land in this area. It looks like an exceptionally good honey plant that would bloom when it is most needed.

Answer - Bee-Bee tree is the common name given to the Evodis Daniellii and it is reported to be an excellent honey plant. The seed should appear where the bloom cluster was but it is our understanding that the seeds do not stand storage well. It might be better to plant the seeds this fall in a well-prepared seed bed and let nature take care of them. Additional help in such a replanting project might be obtained from Fred W. Schwoebel, Langstroth Bee Garden, The Morris Arboretum, Philadelphia, Pennsylvania.

HONEY LOCUST – BLACK LOCUST

Question - I have ordered some black locust seedlings to plant near the bees. Do you know whether the honey locust yields nectar and whether it is preferable to the black locust? I believe that I read sometime ago that honey locust was a misnomer and really yields no nectar at all.

Answer - You were right in your memory concerning honey locust. This is a misnomer. Black locust, in favorable spring seasons, yields quite heavily. However, the honey locust does not yield nectar at all, at least, not in our locality.

HONEY PLANTS

Question - I want to plant some honey plants and trees around my apiary and would like to have your recommendations on what plants and trees would be best.

Answer - The following plants fill in with nectar and pollen for the bees at a time when the major honey plants are not in bloom.

The pussywillows are excellent for providing pollen and nectar very early in the spring. The French variety has larger flowers than do the common varieties. If your beeyard is near a moist area or a stream, purple lossestrife does well under these conditions. The plants start blooming here in late June and the bees work it continuously until October or when frost kills it. It is a hardy perennial and will easily reseed itself. The plant produces numerous spikes on which the flora starts to open at the bottom and continue up as the flora opens.

Chivirico is another plant which produces flora on numerous spikes. The bees work it readily from the time it starts to bloom in August on until late fall.

Vitex is a small shrub which kills back each winter here, but the new growth in the spring produces many tiny blue lavender blossoms. This plant is in bloom from mid-summer to the end of fall.

A couple of trees which are worthy of mention are the Russian olive tree which blooms in early June. It has a delightful fragrance and the bees work it readily. It is used quite often for windbreaks as well as for specimen trees. The other is the Bee Bee Tree. This is a medium sized tree which has large clusters of small whitish flowers during the latter part of July and the first part of August.

ARGENTINE RAPE – WHITE DUTCH CLOVER

Question - I am interested in planting Argentine rape and perhaps a clover (white?) to supplement my annual honey crop. How much soil preparation, what time should I plant and how, and must I add fertilizer or chemical to the soil to attain success?

Answer - Both Argentine rape and white Dutch clover should be sown on land previously disced and harrowed, either drilled or broadcast and then covered by a light harrowing to not cover the seed too deep. The seed of white Dutch clover which is finer or smaller on some soils may need some firming of the seed bed by use of a roller.

The best time for planting white Dutch clover is in early spring. For Argentine rape for bee pasture, we would recommend making successive plantings at 3-week intervals starting in April. Argentine rape may vary considerably in nectar yields in different localities.

Fertilization needed depends entirely on soil where you wish to plant; in most cases it would not be necessary. White Dutch clover is a legume and seed should be inoculated (bacteria culture) unless the nitrogen-fixing organisms are present in the soil where planted.

HONEY PLANTS

Question - I would like to know if there is anything I can plant (Tennessee) from which my bees would make honey.

Answer - It would be difficult to tell you of all the plants that produce nectar but Frank Pellett's excellent book, American Honey Plants, is available from Dadant & Sons, Inc. and describes every plant in North American that is known to produce either nectar or pollen. Almost all plants which bloom will attract bees. The plants in Tennessee that would be attractive to bee would probably include: basswood, locust, tulip poplar, persimmon, maple, willows, sourwood, sumac, berries, the legumes (clover, vetch, alfalfa) and many others.

Ordinarily it does not pay to plant crops solely for the production of honey. You must harvest seed, hay, or something else from the plant to make it pay. The bee-keeper has always depended upon the forest and crops planted by farmers and fruit growers to provide him with his sources of nectar. The bees repay the farmer well with better crops as the result of better pollination.

Question - I have at present 7 colonies of bees. I have recently inherited a 70 acre farm in Newago County, Michigan. Most of the land is cleared and has been cultivated. Part of it is second growth oak and maple. I would like to plant some trees and shrubs for permanent bee pasture. I have been considering planting 100 Russian olive, 100 Siberian pea, 100 little leaf linden and 100 honey locust trees. By planting seedlings, how soon would any of these become a honey source?

I would like to expand my apiary as I grow older and learn more about bees. I hope to plant about 3 acres of sweet clover next spring.

I do not expect to ever use this land for regular farming, but rather as a place to retire with my bees. Do you have any suggestions or advice?

Answer - With regard to the farm land that you wish to plant with bee pasturage, we do not feel that the Russian olive or Siberian pea trees will offer your bees much pasturage. Of course the basswood trees that you mention are an excellent honey plant. however, they cannot be depended upon every year. Also the black locust trees are desirable. You mention honey locust in your letter, but we would like to point out that honey locust is a misnamed tree and does not yield honey. The black locust tree, on the other hand, grows very rapidly, blooms when it is only two or three years old and is an excellent source of very early spring nectar.

Question - I have a round bottom ditch about 150 yards long and 15 feet across. Water stands in the bottom during rainy spells. I'd like to put this land to use. The weeds grow as high as your head in a short time. Should I put honey trees on the shoulder? How long would it take them to bear? Is there a chemical I could use to kill the weeds until I get a new growth of plants started?

Answer - Two generally good honey plants with which we are familiar, Bee Bee Tree, small medium sized tree, and Vitex Negundo incisa, shrub or small tree, should grow in your section though perhaps not in all situations. If the shoulder where they would be planted is several feet above the water level when water stands there, it would seem worth planting a few on a trial basis if you can find a way to keep the weeds out until the plants are big enough to hold their own. Vitex blooms at an early age, usually first year; Bee Bee Tree, after several years.

Also, we would bring to your attention: Purple loosestrife, which in some places spreads wild along waters' edge, in wet soil and may grow to six feet tall. We do not have much information as to how this plant does in the South but it is generally a good bee plant. Chivirico stands up well in competition with many kinds of weeds. Other possibilities: willow trees or shrubs which are generally easy to grow. Willows often yield some nectar and pollen for bees in early spring. Spanish needles grow as weeds in low lands along streams. Some varieties and in some localities are good honey plants. You could try it in your section by collecting seed and scattering it.

Most sprays used in killing weeds would also be detrimental to the other plants. For annual weeds there are herbicides which kill the germination before the weeds come up, if applied before weed growth starts in the spring. These also should be used with some caution lest the material be detrimental to root growth of newly set plants.

Question - I am a side line beekeeper located in southern Ashtabula County, Ohio. Our ground is very hilly and extremely coarse. Alfalfa grows effortlessly and profusely here in this area. This is one of the few places in this part of the country where we can obtain a small surplus of good alfalfa honey.

In addition to our abundance of fall wildflowers we are looking for a fairly dependable source of flora for summer bee pasturage. I am interested in planting sweet clover for the bees only. I have read much about sweet clovers but am perplexed as to which variety or type I should plant to obtain the longest period of pasturage for the bees, plus the most dependable type as a nectar source.

I can obtain white and yellow biennial and Hubam sweet clovers. Do you suppose this soil would support good stands of sweet clover? It grows well here on the side of railroad tracks and in deserted gravel pits. However, where I spread it on waste ground it never took. I have about thirty-five black colonies on the family farm here, and I wondered if five acres of biennial sweet clover, or two and one-half acres of annual and two and one-half acres of biennial would yield enough nectar to give a good surplus of sweet clover honey for this many colonies. I would like any information as to what and how much to plant and when to plant the sweet clovers.

Answer - It is a little hard to answer your question about sweet clover from this distance for your particular locale. We can pass on to you some general information that has been collected regarding the sweet clovers. In the first place, the preferred crop would be the white or the yellow biennial type of clover. Perhaps both should be in the picture, since the yellow blooms first and then is followed by the white sweet clover. Of course, this would result in a longer honeyflow spreading over some 8 weeks of time rather than the shorter flow that you would get from planting only one of the clovers.

From what you say about your country, the reason it grows well on the side of railroad tracks and in deserted gravel pits is simply because there is more lime in the soil in those areas. Sweet clover must have a lime soil to do well and, when it does grow on an acid type soil, there is seldom any nectar in the bloom. We would certainly suggest that you contact your local county agent who will be able to give you some valuable information concerning the soils in your particular area. He also has facilities to test the soil and to tell you how much lime must be added per acre in order to "sweeten" the soil. Generally speaking, it takes more lime per acre to produce nectar in sweet clover than is usually recommended by the farm and agricultural experts.

Five acres of the yellow and white sweet clover would supply a nice addition of nectar for your 35 colonies if the soil were properly prepared and sweetened with lime. That seems to be about the best we can do from the standpoint of recommendation, and we are sure that the sweet clovers would be your best plant to use.

VITEX

Question - I have two shrubs, vitex (vitex incisa negundo). They are excellent nectar producing shrubs and bees work them heavily. They produce lots of seed that are small, round and very hard, but I have had trouble getting them to sprout.

Answer - Melvin Pellett of Pellett Gardens, Atlantic, Iowa stated that he has never failed to get a stand of vitex from seed but sometimes some stands are better than others.

The seed is planted in early spring in well cultivated soil that is as weed free as possible. The seeds are placed 1 to 2 inches in the ground and the seed bed is kept well sprinkled until they sprout and the plants begin to mature. The Pellett Gardens use an irrigation sprinkler system which gives a slow, steady flow of water for moisture purposes. The statement was made that if a heavy watering was done, the soil would crust and this would inhibit the plants sprouting and growing. It is necessary to keep the bed weeded until the plants get a good start. Apparently vitex is not real easy to grow but it is entirely possible.

FALL HONEY PLANTS

Question - This fall as I walked near my colonies, I noticed a very strong odor coming from the hives and it wasn't necessary to open the hives to smell it. Could this be disease?

Answer - *Both AFB and EFB, in heavy infections, do emit odors that can be detected by walking somewhere near the colonies. It would certainly be a good idea to check for disease with a thorough examination of the colonies. We would suspect, however, that the aroma you noticed was coming from newly gathered fall honey. A couple of common fall honey plants in your area (Illinois) are heartsease (smartweed) and Spanish needle. Both yield a dark colored, highly flavored honey. Some people find the honey flavor from these sources objectionable but many people do like it. Bakers frequently prefer the heavily flavored fall honeys since they impart their aroma to the baked goods. The honey from these fall sources makes excellent winter stores for the bees.*

GOLDENROD

BUCKWHEAT

SPANISH NEEDLE

Question - One of my colonies in a three-week period of time drew out and capped nine frames. Some of this came from heartsease and some from a plant about four or five feet tall that has bright yellow-petaled flowers something like a daisy. Would this plant be what they refer to in bee journals as an aster? Also, I have some golden-rod in our vicinity (Iowa). Does this ever produce some nectar or a surplus in my location?

Answer   - The bright yellow-petaled flowers are probably Spanish needle (Bidens aristosa). The florets of the fall asters which produce most of the nectar are white. Although goldenrod is quite prolific in eastern Iowa, generally speaking it should be considered a minor source of nectar.

BUCKWHEAT

Question - Last year I planted a patch of buckwheat (Illinois) which bloomed well all season, but the bees did not work on it. I did not see one drop of buckwheat honey. Why?

Answer - Buckwheat is a plant which does rather poorly in our part of the country. It does best in the states of New York, Pennsylvania, Ohio, Wisconsin, and Minnesota. Buckwheat has the reputation as being a crop to grow on poor ground. Many years ago, though, there was an area near here where the farmers would work up the ground each summer and plant buckwheat. At that time the bees in that area would make a fair amount of honey but it was of rather poor quality and quite dark in color.

HAIRY VETCH OR HUBAM CLOVER

Question - Do you think it would pay if I bought seed (Hairy Vetch), for 30 acres of unused soil that is behind me and let the owner plant the seed and harvest it, and I take whatever the bees make? How many hives could work it and what is the crop average for Hairy Vetch in this part of Texas (Mesquite)? If you think this could be to my advantage but think another crop would be better, I am open for any suggestions. I did not suggest clover because it is relatively expensive compared to vetch. Hubam clover does well here because there is about 3 acres not too far away that does fairly well on nectar production.

Answer - Hairy vetch is an excellent honey plant down there, but at the same time, this is also true of Hubam clover. May we suggest that you contact Mr. Curtis Meier, Manager at Dadant & Sons, Inc., 1169 Bonham St., P.O. Box 146, Paris, Texas 75460. In addition to being branch manager, Mr. Meier is a commercial beekeeper in Texas and will be able to give you knowledgeable answers relative to honey plants in the state of Texas.

CLOVER FOR HONEY PRODUCTION

Question - I would like to plant some clover this fall for my bees. Which is best, white sweet clover or Madrid clover? Will the Madrid clover bloom the first year? Which is best for my climate?

Answer - We do not think that you can afford to plant crops just for honey production. However, Madrid clover is an annual sweet clover and will bloom the first year. White sweet clover is a biennial and will bloom the second year. Either of the clovers is suitable for your climate (Dallas, Texas), but we would suggest that you contact your local County Agricultural Agent or State University to get their reaction to the desirability of these plants.

SNOW-ON-THE-MOUNTAIN

Question - My bees are bringing in honey right now that has a reddish color and has a weed taste and odor. It is edible but not top quality honey. I'm wondering what plant or weed the bees got it from. Right now we have in bloom (Oklahoma): ragweed, snow-on-the-mountain, broomweed (August flower), acres of a silver-gray weed which grows 3 to 4 feet tall with flowers on limbs on top, tamarisk, goldenrod, a yellow aster (looks like a daisy bloom), sunflower (acres of them). There are also watermelons near the bees to get juice from. I don't know of any smartweed they are close to but the honey has a smartweed odor. Also mimosa trees are in bloom. Hope you can give me a clue on this.

Answer - We went to American Honey Plants by Frank C. Pellett for an answer to this one and snow-on-the-mountain produces a strange flavored honey that has a red color. It is the only plant on your list, that is reported to produce honey which is red. Tamarisk produces a low grade, dark honey with a minty flavor according to Pellett.

APPLES IN THE SPRING

CHAPTER 9

HONEY

A FRAME OF BEAUTIFULLY CAPPED HONEY

Question - I was wondering if honey that had been overheated and burned could be fed back to the bees?

Glenn Booth
Austin, MN

Answer - The quick short answer is yes, if you do it when the bees can consistently get outside to defecate.

The longer answer as to why it should be done only under the above conditions is this:

If you or anyone you know has ever made candy or made a sugar glaze when cooking, one of the things you don't want to do is get this sugar material too hot or it will burn or scorch. What happens when you burn sugar is that it is reduced to its carbon base. Carbon is black, tastes icky and is indigestible to you and to honey bees.

Pure sugar or sugar syrup has no indigestible material, no extraneous matter, so nothing is left in the intestine after digestion and absorption to be excreted. But, if you have burnt sugar/carbon, the carbon is left and has to be excreted. This is OK if the bees can get out to do it. If it is fall, winter, extended rainy periods, etc., where they can't, a problem may exist as you can see.

So, go ahead with the above disclaimer in mind.

Question - Is it advisable to melt granulated honey in a microwave oven? Where can I find large microwave ovens?

Enrique Ulibarri O.
Morrlia, Mexico

Answer - Granulated honey can be melted in a microwave oven. I've done it at times in my own home with small quantities of honey. As in trying to reliquefy honey using any heat source, you have to be careful of the amount, intensity and duration of the heat applied. The honey can be darkened and burnt easily if this is not carefully watched. I am not aware of large microwave ovens that can be used to liquefy commercial quantities of honey. I'm sure that the size necessary may exist. You will have to contact companies which supply equipment for food preparation in restaurants, schools, the military, etc.

As most of those who use microwave ovens know, the heat is uneven, so hot spots and cold spots will occur. This is not as bad with liquids, however. Also, many microwave ovens now compensate by rotating the food as it heats.

COMMENT

Dear Jerry:

Regarding your reply to the question about solidly crystallized combs in "The Classroom" column in August ABJ, here is another very satisfactory way of dealing with the problem, without destroying the combs. First, thoroughly scratch the surface of the combs to remove any cappings and then stand them in cold water. A few can go in a sink - a lot in a barrel. It usually takes overnight for all the honey to be dissolved and then the combs can be shaken to get the dilute honey out. The combs are as good as new and the dilute honey can be used for a variety of purposes: for example, mead making or add extra sugar for winter feed. Nothing is wasted!

David Dawson
Manitoba, Canada

MOISTURE IN HONEY

Question - I know honey isn't ripe until all the cells are capped after conversion of nectar to honey involving chemical changes brought about by enzymes and physical changes due to evaporation of some of the water contained in nectar. So, other than this are there any environmental problems that may delay the ripening process after the chemical changes from nectar to honey?

Linda Stamper
Ft. Lauderdale, FL

Answer - The only environmental problem to the proper ripening of nectar to honey would be high humidity. In areas of the country with high humidity (like the southeastern U.S.), sometimes the bees have a difficult time getting the moisture level below 18.5%. This is the key level as anything above and honey may ferment in the comb. Because the capping process is not instantaneous, the honey can pick up additional moisture before being fully capped in extremely high heat/humidity parts of the country. The honey also may pick up moisture after it is capped if the humidity is too high. By the same token, the reverse (low humidity) can eliminate unwanted moisture in capped honey. That's why some beekeepers use "hot rooms" in their honey house equipped with dehumidifiers.

Some commercial beekeepers own honey dryers which drive off moisture with heat in a vacuum. The vacuum allows the excess moisture in liquid honey to actually be boiled off since water boils at a much lower temperature in a vacuum. The lower temperature is necessary to prevent heat degradation of the honey.

Question - With freeze drying used in so many different kinds of food, why isn't there dried honey?

Evelyn Hale
Nauvoo, IL

Answer - Good Question. For a good answer I went to Sherri Jennings of the National Honey Board in Longmont, CO. Here's her answer:

"I asked our food technologist why there is not a dehydrated honey product available that is 100 percent honey. I was told that because honey is a humectant and picks up moisture, it won't free flow like table sugar without carriers. The carriers (such as starch) are used to prevent the product from clumping.

With the advent of new technologies such as microwave drying, this constraint may soon be overcome. The Dried Foods Technology Laboratory at California State University, Fresno, is the only facility in the United States suited for the demonstration of the microwave - vacuum (MIVAC) dehydration technology. Using this technology, researchers have been able to dry honey successfully. The Honey Board was considering approving a research project to explore the possibilities for a 100 percent dry honey product. The product would either be formulated in a desirable end-product (i.e.: throat lozenges, chocolate or cosmetics) or ground to obtain a pure honey powder."

EXTRACTING HONEY

Question - Before the advent of modern bee hives in the early 1900's, bees were kept as early as Egyptian times.

My question to you is how was the honey extracted from the natural made bees nest? How did they get the honey out?

Victor Landry
Windham, NH

Answer - In many parts of the world today honey bees are still kept in hives that do not have movable frames. In other words, the combs are attached to the container used as a hive. The combs that have honey in them are simply cut out and removed from the hive.

The combs are crushed and placed in some type of large vessel and the honey and wax separated naturally, with the wax floating on top of the mostly liquid honey. Some straining may take place to produce an even cleaner honey if that is desired.

This is very low-tech as you can imagine, but is effective and has been done for thousands of years.

Question - Late in the season last year I discovered a large area - about 10 acres that remains under water and was covered in flowering yellow lotus plants.

I am wondering if you can advise me as to whether or not this may be a good nectar source and if it would be of good flavor.

Doug Bem
Muskogee, OH

Answer - I'm going to assume that your Lotus flowers are the same as our Lotus or Water Lily flowers that we have on the Mississippi river here in western Illinois.

Several years ago I placed some hives in a location close to about 80 acres of water lilies. Those hives collected a lot of pollen and a very "fruity" honey from those blossoms. Five colonies only produced about a super apiece from this source, so I'll have to guess that nectar was not available in great quantities. Those colonies did collect lots and lots of pollen and the hives did very well.

I'd give the location a try, but maybe use only a small number of colonies and see how it goes.

Question - Can you tell me exactly what honey consists of?

Phillip Glosick
Bath, New York

Answer - Well, nothing is exact about honey because of different plant/nectar sources, but I can get close. Here goes. Potassium, sodium, calcium, magnesium, iron, copper, manganese, chlorine, phosphorous, sulphur, silicon, chromium, lithium, nickel, lead, tin, zinc, osmium, beryllium, vanadium, zirconium, silver, barium, gallium, gold, germanium, strontium, water, fructose, levulose, dextrose, sucrose, maltose, gluconic acid, lysine, histidine, arginine, aspartic acid, threonine, serine, glutamic acid, proline, glycine, alanine, cystine, valine, methione, isoleucine, leucine, tyrosine, phenylalanine, tryptophan and a bunch of enzymes, flavors and aromas.

How's that for being analytical!

CRYSTALLIZED HONEY IN COMBS

Question - I have some supers of crystallized honey. What is the best way to get it out of the comb?

Charles Damron
Wyandotte, MI

Answer - About the only thing you can do is put these supers on colonies that need food and let them clean them out. Sometimes this may not work either if the honey is crystallized very solidly. At times like these, the whole comb must be melted to separate the wax from the honey.

BEAUTIFUL HONEY GIFT PACK

Question - The article on the therapeutic benefits of honey because of its bacterial properties has led to some confusion about the properties of honey. The article stated that honey acts as an antibiotic by dehydrating bacteria, therefore, honey acts as a good dressing for burns and wounds.

Honey must be a carrier or some diseases since I have heard that you should not feed honey of unknown origin to healthy honey bees for fear of infecting the bees with 'American Foulbrood' or other diseases.

It has also been said that honey should not be fed to infants because of the fear of botulism.

How can honey be a cure for some diseases and yet be a carrier of others?

Gary Hawkins
Lawrenceville, GA

Answer - It is confusing when the various life stages of some bacteria are so different from what we understand to be life.

The short answer is that under harsh conditions, heat, cold, dryness etc., some bacteria change into what we call a spore form. They very quickly develop a hard outer covering and cease life as we know it. They can remain in this form for decades. Then, when conditions are right, they change back into the original bacteria.

Honey of 18% moisture or less is a harsh environment. Bacteria cannot live in it. But the spores of these bacteria as we spoke of above can be present waiting for conditions to get better and change into a functioning organism. When honey is diluted above 18% moisture or when another bee eats the honey and it's diluted and changed within its digestive system or when an infant who consumes honey and who is possibly constipated allows the spore form to find a suitable growing environment, then watch out because the bacteria is now alive!

The bacteria is looking for a home to grow and reproduce. When no honey is available, they become a dormant non-infective spore. When a home with warmth, moisture and other factors is found, the spore turns back to a bacteria which may harm other life forms.

Question - Will honey ferment faster in plastic containers than it will in glass containers?

Cindy
Mead, Washington

Answer - Honey fermentation is based mainly on the percentage of moisture in the honey and the natural yeasts which are present. Honey with moisture below 18% generally will not ferment. The low moisture content will not allow the yeasts to start using the sugar, start multiplying and start producing carbon dioxide and alcohol. Honey having moisture above 18% is a better environment for fermentation. It has nothing to do with using glass or plastic containers or lids. Plastic will not allow moisture to migrate into your honey. Be sure that lids are on tight and you'll be fine as long as the moisture is low enough.

You could heat your honey to 125º F for 470 minutes and kill all the yeast like the commercial packers do, but I've always liked a more natural product which can be obtained with a little more care and attention. Also, any heat is detrimental to delicate honey flavors and aromas, not to mention the darkening effect on honey color.

Question - I am working on constructing a small commercial honey house. I have access to some cheap aluminum pipes and wondered if this would be acceptable for pumping honey from the extractor to other tanks.

Tracy Turner
Cadillac, MI

Answer - No, not a good plan no matter how inexpensive it is! I don't know if you have ever seen aluminum that has corroded, but it's the same process (oxidation) which results in iron rusting, only a different cause. Honey is a slightly acid product and this acid will cause the aluminum to start to corrode. When aluminum corrodes, the aluminum changes to a different substance, aluminum oxide, which sloughs of the aluminum and would be mixed with the honey. Too much corrosion and the pipe could break or fail in some way. The worst aspect though is the aluminum which is now in the honey. Someone consuming too much aluminum either in the food they eat or water they drink or air they breathe, can suffer major health problems. In fact, scientists have discovered that the people suffering from alzheimer's disease have elevated levels of aluminum in their brains. So, do not use aluminum pipes in your honey house and go into your kitchen and see how many aluminum pots and pans you have that are used for cooking.

Question - Recently a buyer called trying to buy honey. He had bought some at a vegetable market and was disappointed with it because it had become solid rather than free flowing from the jar. I know that putting it out in the sun will liquefy it, but would like to know if there is a reason it becomes solid in only four months. Thank you for any information you could give on this subject.

Betty Turner
N. Charleston, SC

Answer - Most, but not all honeys, will granulate or crystallize. Some may take months or years, others just weeks or days, even granulating in the comb. The tendency for granulation is related to the sugar composition of the honey its moisture content and the temperature.

As you know, there are many different kinds of sugar in the nectar that bees gather from flowers. Most of the time a honey that will granulate the quickest has a higher dextrose content in relation to the other sugars present. For instance, the honey from Canola, of which thousands of acres grow in Canada and Europe, will granulate in the comb and cause all sorts of management problems for beekeepers. I've had pure Black Locust honey in my cool basement for over a year and is hasn't given any indication that it is granulating.

It all depends on what the sugar ratio is between the dextrose and the other sugars in the honey. This is variable, depending on what flowers were foraged on.

As an aside, we have been talking like granulation is a bad thing. In most of the rest of the world, the consumer prefers a finely granulated smooth creamy honey. It's easier to use. It does not run or drip or make a mess when using it. I think this is a market niche for honey that hasn't been fully tapped as yet in the United States.

POLARISCOPE

Question - In the September issue of the *American Bee Journal*, "Grading Your Honey" was an interesting article. We would like to find the materials with which to make the polariscope.

Can you tell us where we might find the polarizing film mentioned in the article about the polariscope, or the company to which we can write?

Gail and Carl Chieffo
Girard, OH

Answer - There is an excellent catalog of thousands of science-related items for amateurs to professionals put out by Edmond Scientific. They have several pages of Polarizing films, disks, etc. Call them for a catalog at 609-573-6250 or 609-547-3488. Besides the polarizing film, you will really enjoy the catalog.

HONEY CRYSTALLIZATION

Question - I am in my first year with honey bees, and am thrilled with the experience and with the finished product I obtained from the bees and hive your company supplied me with.

I have one small problem, however, and that is the honey crystallization within a few days of extraction. I put it in small plastic containers and kept it inside the house, and within 3 days I could not pour the honey out.

I was told you may be able to tell me why it would do so, and possibly how to prevent recurrence in the future. I am looking forward to a long future with my bees and hope to be a long customer of Dadant & Sons.

John L. Scripps
Casnovia, MI

Answer - First, honey that crystallizes is a normal, natural occurrence. It is a sign that the honey has not been treated in a way, which I'll explain later, changes it. Honey that crystallizes is used as a sales marketing tool in some places to prove the naturalness of the product.

Honey crystallizes because of the proportion of certain sugars in the honey which enhance those sugars' ability to form crystals. Some honeys do not crystallize at all and some like Canola crystallize so quickly that the beekeeper must remove it from the supers within days of the bees storing it in the supers.

If you have ever seen or tasted "creamed" honey, which is finely crystallized honey, you would know why the majority of the world eats it. But, that wasn't your question.

If you heat your honey quickly to about 130-140ºF and cool it quickly to 90ºF or so, you can retard crystallization. This heat liquifies any existing sugar crystals and slows the process. If the heating and cooling is done quickly, the honey will not darken as much. you will lose some aroma and taste and destroy some enzymes. That is the trade off for liquid honey with which many commercial honey packers must live.

THICK HONEY

Question - I do have a unique problem or so I think. I need advice. This is the first year for me here in Nevada with my bee hives. I have two colonies and I have them in the area of Carson City, Nev.

I have had hives in Utah and California in the past. I have had no problems in extracting the honey after the flow before. But, extracting is a real problem here.

We live in a semi-arid area, and the only waterer that is close by for the hives is my large garden and the horse waterers on the adjacent property. When I tried to extract the honey that I placed in the extractor just like I have done in the past, no matter how hard I spun the frames, the honey would not flow. I took a capping knife and cut into the frame and the honey is so thick it just stayed on the end of the knife and did not drop until I placed a heat torch to it.

My question is this: Just how do I extract the honey from the hives? I have placed the frames close to the garage door, which received sun and warmth all day, and I have taken the extractor into the kitchen and I still do not get any flow.

If I cut out the comb, and get a solar melter, or extractor how do I separate the honey from the wax?

Please advise.

Toby Clark
Minden, Nev.

Answer - I lived in southern Nevada for several years and kept bees and got some beautiful catsclaw acacia honey many times, but never had the problems that you are having. But when relative humidity in the desert averages about 10% or lower, I can visualize some ripened nectar being pretty thick. However, be sure your problem is really thick honey and not simply granulated honey in combs.

Let me give you a short list of methods and alternatives in removing the honey from the combs.

1) Try to warm the honey in the comb to about 90°-$100^{\circ}F$ and extract slowly as the wax will have softened slightly.

2) Melted beeswax and honey will separate. This is the principle used in cappings wax melters and may work here. Beeswax melts at $150^{\circ}F$ and will float on top of the honey in the wax melter. Remember, honey discolors and loses some of its taste and aroma when heated at too high a temperature and for too long a period of time. Also, the beeswax is highly flammable and should not be heated directly.

3) In some countries that have thixotropic honey, which is honey which has the property of a gel-like consistency when in the comb, the use of presses to squeeze the honey out of the comb are employed widely.

4) Perhaps next year, if you haven't already, use cutcomb or thin surplus foundation and produce comb honey.

As you can see, there is no neat, clean answer if you want to produce liquid honey. Perhaps some of the above ideas will help. I would like to know how you solved this problem, so please drop me a note when you can.

Question - I've been keeping bees several years now. Every year I put back a hundred pounds or so of honey for myself to use through the year. My question is, how long can you keep honey and it will still be good to eat?

Thomas Mulinix
Kingston, GA

Answer - As long as the moisture level is kept below 18% to inhibit fermentation, honey will "keep" just about forever.

Honey that is still edible has been found in pyramids in the tombs of nobility. It was thousands of years old. Honey does darken over time and its flavor lessens also, but its sugar composition never changes if its moisture content stays low.

Hopefully, though, your honey is so good and well bottled that it won't last until the next season. If it does last, it won't be as good as when you first extracted, but it will still be better than all the other sweeteners in the stores.

If you have room in your freezer, consider storing your honey there. Frozen honey is almost indistinguishable from fresh honey. Very little degradation takes place.

Question - It seems like a dried, powdered honey would be a wonderful product for consumer and industrial use. Why hasn't anyone done this?

Andrew Rivera
Lima, Peru

Answer - *It has been done and it can be done, but as of this date, no company in the U.S. we could find is currently making or distributing a dried honey product. Two products were developed: The first was a baker's formulation consisting of dried honey and starch. The second product was actually pure dried honey which was marketed in a "sugar-like" consistency. I had heard that there were some technical and marketing problems that were not overcome and therefore the product was dropped. For one thing, dried honey is extremely hygroscopic (water absorbing), so one must keep it in sealed containers or in a low humidity environment.*

I personally think a dried honey could be successfully marketed. Like everything else in life, it takes the right person at the right time with enough money to make things happen. Maybe that person is you. Good luck to you!

Question - The Superior Land Beekeepers' Association is interested in learning how to dry honey into a powder form. We will appreciate your help in this matter.

Maxine Smith
Daftee, MI

Answer - *There is not a whole lot of information on drying honey. Not that it's difficult to dry, but it is difficult to keep dry. Once dry it wants to grab on to any moisture in the air and it becomes a hard candy-like substance. You would think that with all the super duper moisture barrier/foil packaging available that this would be a no-brainer by now. But, apparently it isn't viable from a packaging standpoint or a marketing view. Anyway, it still sounds like a good idea to me.*

What follows is a short discussion of dried honey from the book Honey edited by Eva Crane.

DRIED HONEY

"Dried honey, or honey concentrate, is made by spraying a film of honey on to a slowly revolving hot drum about 4 m long and 2 1/2 m in diameter. At the completion of each revolution the film of honey, from which the water has been evaporated, is scraped off. Dried honey can also be produces by passing it through an evaporator under vacuum (Turkit et al. 1960, Chaffey et al., 1961). When the pure honey

concentrate is exposed to the air, it soon takes up moisture and sets into a hard and unmanageable mass. A practicable product can, however, be produced by blending the concentrate with about 55% of a starch of a non-hydroscopic sugar, or a combination of both (Glabe et al, 1963, 1965, Shookhoff, 1957, also a Japanese patent, Tokyo Yakuhin Kaihatsu, 1969). The stable powder thus produced is used in dry mixes for cakes and bread; it improves their flavor and texture, without the necessity of handling or packaging the honey as such (Food Processing, 1971). These types of honey products have been developed to suit modern manufacturing and marketing methods. At present the world consumption of honey in them is not more than 50 tons a year, but this amount will undoubtedly increase, and they may become one of the important commercial honey products."

WHAT YOU'VE BEEN WAITING FOR —
A FULL SUPER!

Question - Several years ago I began putting herbs in honey and the result was a very tasty honey which I gave away as presents. Everyone loved them so well I would like to start marketing them. My question is, what are laws about doing this? What I do is heat the honey just a little and then put the fresh picked herbs in the honey – herbs such as lemon balm, basil, peppermint, spearmint, anise hyssop and so on. The honey is just a little thinner but delicious. I have several jars which I have had for about five years and they are just as delicious as the first year I bottled them.

Theresa Martin
Farmington, ME

Answer - *It's exciting to hear when someone is being creative when working with honey sales. I hope you can carve out that niche for your honey product.*

Every state has similar but different food preparation, packaging and serving rules and regulations. They do this so that reasonably sanitary and properly prepared and packaged food reaches the consumer.

This is one area where one has little luck "fighting" City Hall. You are doing the right thing by doing your research and finding out the laws that will allow you to sell a food product to the public.

I would contact Tony Jadczak, Maine Dept. of Agriculture, Division of Plant Industry, Augusta, ME 04333, Ph. 207-289-3891. He is the contact for the beekeeping industry in Maine and he will know exactly which government offices you need to contact.

There shouldn't be a problem with selling your product as long as proper bottling and labeling laws are followed. Similar products have been sold for many years by other beekeepers and packers.

Question - In the fall, active bee hives have a strong musty smell, even from several feet away! At this moment my nearby window hive has this smell. What causes this smell?

When I extract honey I find that some honey frames cannot be extracted because the honey had crystallized, left over after wintering in the supers, no doubt. Bees don't seem to like crystallized honey in the frames, so what does one do usefully with those frames?

Fred Fulton
Montgomery, AL

Answer - *The strong musty smell is usually nectar ripening that the bees have collected from fall flowers. I've had nectar ripening that I knew the bees had collected from goldenrod and asters that smelled like dirty socks. I like the honey, but the ripening part was stinky.*

You are right, bees don't like crystallized honey. It takes time, effort and water to reduce the crystals to a sugar solution they can use. If you have frames and frames of it, sometimes it can be washed out using a spray nozzle on a garden hose at low pressure. One beekeeper also indicated he has had luck solving the problems by soaking the crystallized frames of honey in large barrels of water. The water loosened up the crystals enough so that the bees could finish cleaning them.

You can, of course, cut everything out of the frame and render the wax out throwing everything else away. Or save it and use it early spring to feed starving colonies or package bees or splits you've started, they'll take it out.

Question - I've managed to pull off an early crop of very light and delicious honey. With the county and state fair honey contests coming up, I have visions of blue ribbons. In order to get entered in the proper classification, I want to be clear on color grading. Any ideas on how to grade without a Pfund color grader or a USDA color comparator? Also, do you have any tips on final filtering on a hobby scale?

M. Ware
Gilbert, Colorado

Answer - I'm glad you have gotten such high quality honey.

For the hobby beekeeper, color grading and testing for moisture levels can be the most difficult because of availability of the test equipment needed.

Without a USDA color comparator, a Pfund honey grader or the newer Lovibond honey grader, you can only guess at the color class. Without any great prior experience in color grading, your guess would be just that, a guess. You may want to contact the president of your local beekeeping association and ask who may have color grading equipment that you could access. While you're asking about this equipment, ask about who has a refractometer to test the moisture level in the honey. You can either lose or gain points from judges on the moisture level.

To remove much of the "turbidity" in your honey, it first needs to sit at room temperature 70ºF - 80ºF to allow bubbles and other extraneous material to float to the top where it can be skimmed off. If you decide that additional filtering is needed, a low tech filter medium is a clean nylon stocking. They are lint free and do a good job of filtering out small material. My wife has donated stockings in the past to me sometimes without her donation being known until after the fact. Good thing they are fairly inexpensive.

Have fun with this competition. You'll learn more this year that you can apply and use next year to do an even better job. Good Luck!

One additional suggestion is to obtain a copy of the previously used judging score cards. This will tell you exactly how the judges are deciding which entries are best.

Question - To give you an idea of our climate, my town in Arequipa in Peru is situated a 7,500 feet above sea level with a sunny and very dry climate around 10 months of the year and with an average daily temperature of 75ºF. During the winter, the night temperature drops to around 32º - 35ºF. The relative humidity is around 8% (except in our rainy season, January & February).

Yesterday I took out eight frames of honey from one of my hives and I noted that the honey was extremely thick (a very small portion in each frame had crystallized). As it was too thick to extract in the honey extractor, I scraped it out of the frames, and put it in a container in "bano maria" (a hot water bath) to separate the honey from the wax.

My question is: is this honey as good as that I normally extract, and if I find more like this in the other hives, should I give it the same treatment? Or, is there a reason why this honey should not be used?

Anthony Gomez
Arequipa, Peru

Answer - I lived for many years in our state of Nevada. It was very dry there also. Humidity ranged at the 5 to 10% level. The rainy season was in January and February there as well. The few colonies of bees I kept there did very well on the desert plants around. There was, surprisingly, plenty of pollen and nectar. I had the same problem you had trying to extract the honey. It was too thick. It just wouldn't come out easily. The humidity was so low that the moisture in the honey would be around 10% or lower, making the honey very thick. The bees had no problem lowering the moisture level in the nectar they brought in. I resorted to just producing comb honey to get around this situation.

The honey separated from the wax in your hot water bath probably would not be as aromatic or flavorful as right out of the comb, simply because of the heat necessary to separate the honey and wax. It would be O.K., but any time a lot of heat is applied to honey, it does some harm to the product.

You could try pressing the honey comb, forcing the honey out of the wax similar to the process of squeezing the juice out of grapes. In some places in Europe they produce honey from Heather. This is normally very thick and has to be pressed out.

Since nature has given you a certain situation perhaps, if possible because of your market or desires, you should not try to conquer or overcome your thick honey in the comb. Use it as an advantage! That may mean selling it as "fresh honey in the comb".

Question - Many people in my community believe that local honey is good for their allergies. I have not noticed any information on this in the *Bee Journal* or other books and literature that I have consulted.

I would really like to know if there is any good evidence that locally produced honey is helpful in relieving allergy symptoms. Any information that you have would be appreciated and of possible help to many people.

Richard Sheridan
Sheffield, AL

Answer - Allergy symptoms, the red watery itchy eyes, cough, flu-like feelings all over your body are miserable to say the least. Unfortunately, most products to treat the symptoms contain antihistamines or steroids that can have side effects which aren't much better than the allergy. No wonder millions of dollars are spent on relief and that new relief methods are sought.

Eating locally honey-bee collected pollen or honey with whatever pollen is suspended in it is sometimes reported to help allergy sufferers. Here are a couple of things to keep in mind. The human body cannot digest all pollen. The pollen grain coating is silica (glass) usually and is designed to protect the reproductive goodies inside against heat, light, moisture, drying, etc., etc. The thought is that your body recognizes the pollen over time and you become "immune" if you will, to it (similar to allergy shots). Well, the body can't recognize the pollen content if we can't digest it. Maybe it's the shape of the pollen the body may recognize. There are two general classes of pollen. One is sticky and has projections and weird shapes that insects col-lect. The other is rounder, smoother and lighter to be able to be blown about by the air. People's allergies are generally caused by windblown smooth, round pollen not sticky, poky, convoluted pollen, so it can't be shape can it?

Like a lot of other folk remedies, sometimes eating local honey or pollen does seem to work (placebo?), sometimes it doesn't. But the medical establishment hasn't looked very hard either.

A British research report I once read claims it's not the pollen, but another ingredient in the honey which helps allergies.

Whatever the reason, it is obvious that this folklore remedy has been around too long to simply dismiss it as foolishness. What is needed are serious <u>scientific</u> studies and <u>doubleblind tests</u> performed by reputable universities or other research groups. Only then can we either endorse or refute the honey/pollen allergy folk medi-cine cure. The problem <u>is</u> who is going to support this kind of research?–surely not the big drug companies since you can't patent honey or pollen!

HONEY FERMENTATION

Question - For the past 15 years or so I have operated a sideline honey business which today amounts to 51 stands of bees in three widely separated yards.

I have had very little trouble with granulation and none with fermentation until this past year when both have become a serious problem.

In the past month or so I have had to pick up several cases of honey that I had sold to stores because it had apparently fermented. Each jar had a heavy layer of foam on top and the honey had forced its way out through the tops of the jars and down the sides. Upon tasting the honey I discovered that it had a very sour taste as though vinegar had been mixed with it.

Answer - We were sorry to hear the problems you have been having with granulated, fermented honey. Fermentation of honey is caused by the action of sugar-tolerant yeasts upon levulose and dextrose, resulting in the formation of alcohol and carbon dioxide. The alcohol in the presence of oxygen then may be broken down into acetic acid and water. As a result, honey that has fermented may have a sour taste such as that which you tasted.

Ordinary yeasts do not cause fermentation of honey because they cannot grow in the higher sugar concentration. Spoilage by bacteria is not possible because of the high acidity of honey. The primary sources of the sugar-tolerant yeasts are the flowers and the soils. It can be assumed that the yeasts which cause the fermentation of honey would be common and present in most if not all honey that is produced.

The chief factors in honey fermentation are yeast and moisture content. Inter-related with these are the storage conditions and presence of granulation. It has been published that honeys with less than 17.1% water will not ferment in a year, no matter what the yeast count may be. If the moisture content lies between 17.1 and 18%, honey with a yeast count of 1000 per gram or less will be safe from fermenta-

tion for one year. For honeys with moisture contents between 18.1 and 19% moisture, a count must be only 10 per gram to guard against fermentation for a year. Above this moisture level, more than one yeast spore per gram means an active danger of fermentation. Granulation of honey always increases the possibility of fermentation because of the appreciable increase in moisture content of the remaining liquid portion.

If honey is heated to 145ºF for 30 minutes, or at higher temperatures for shorter periods of time, it will be safe from fermentation if protected from further yeast contamination. However, the simplest and easiest method for coping with the problem on a small scale would be to be sure that the moisture content is low before bottling the honey. This can be accomplished with a dehumidifier, heater, and fan placed in a small room with the staggered honey supers.

HONEY CONTENTS

Question - Concerning the article page 681 Nov. 1991 of the ABJ "A Visit to a Japanese Honey Packing Plant." If the honey contains 65% sugar, and 21% water, what else is in the honey? 65% plus 21% is 86%. What is the other 14%?

Gerhard Guth
Micanopy, FL 32667

Answer - Many times the percent of sugar only take into consideration the two with the highest concentration, usually fructose and dextrose. If you add in the percentage of sucrose, maltose, and other sugars along with acids, proteins and ash (minerals) you will get the other 14% that wasn't designated.

Question - I am a hobby beekeeper and eat or give away all the honey I harvest. I have always had good-flavored honey and get a lot of compliments on it. This year I had a lot of *bad* honey, and needless to say, am very upset. I have no idea what I did wrong. It has a flat taste and leaves a bad taste in your mouth. One friend described it as an oil taste. One hive had much darker honey than the others, but they are 7 or 8 miles apart.

The only things I did differently this year were that I left the supers stacked in the honey house several weeks before extracting, and I heated the honey to 150º - 160º. I took the honey from the extractor in plastic "Rubber Maid" buckets and set them in hot water to heat. Then I cooled it after bottling. I haven't had any kind of treatment in my hives except Apistan strips last winter. Is there some place I can send a sample to get it checked? Or maybe you have some ideas.

A.L. Kerby
Bovina, TX

Answer - I'm sorry to hear of the problem you are having with your honey. I have some things for you to think about.

There are only a few things which can affect the type and quality of the honey. One is obviously the species of plant that is blooming from which the bees are collecting nectar. Even if your bees are in the same location, year after year, plants that bloom can vary in their quantity and quality of nectar. This is not the beekeeper's fault. It is just the way that nature varies things. Check with other beekeepers in your area to find out if much honeydew was produced by colonies this last season. It is darker and stronger flavored than regular honey.

The things that are the beekeeper's responsibility when handling the honey are the containers that come in contact with the honey and the temperature that the honey is subjected to when extracted and stored.

Always use containers, extractors included, that are "food approved"! What this means is that they won't give off any odor, taste or color to the food that is in contact with them. That is why stainless steel and glass are generally chosen for most food processing, honey included. There are some plastics which are also "food approved", but they are more expensive than most plastic containers you can buy in a discount store. Just because the product has a well known and regarded name does it mean that it is a "food approved" container. A plastic container will have the words "food approved" written on it if it is intended for that use.

The temperature that is applied to the honey during processing is also important for the beekeeper. All honey processing should be accomplished below 100ºF. Anything above will damage the honey. The temperature you spoke of will drive off most of the volatile oils which give honey its taste and aroma, leaving a bland taste that probably has a burned component to it because of the high temperature you subjected it to.

Couple these two very important things together, the container used and the high temperature and you have honey that probably is ruined.

The best lessons learned sometimes are the hardest. You'll do better this year because of last season's experience.

BOTULISM AND HONEY

Question - As a small beekeeper in southern Louisiana, recently I encountered a manager of a local restaurant who was very concerned about the potential problem posed by botulism spores in raw honey. Her assumption was that pasteurized honey did not pose that threat. In trying to respond, I realized I needed more information on this subject myself. Specifically, is the potential for botulism spores really any greater in raw honey than in pasteurized or processed honey? That is, does the heating and filtering of honey actually assist in the destruction and/or removal of the spores? One Louisiana Dept. of Health official recently stated that there was no greater threat of botulism spores from raw honey, but he said he had no information to document that statement.

Do you know of other small beekeepers who have run across this same question in selling their product?

Jay Martin
New Orleans, LA

Answer - *Botulism spores are everywhere – in the air we breath, food we eat, clothes we wear. The spores themselves are not harmful. It is only when the environment is right that they develop into the botulism organism that secretes the toxin (poison) which makes man ill and may cause death in certain cases. Botulism organisms require warm, moist conditions and no oxygen. They cannot grow in the presence of oxygen. This means that there are only a few situations where they can grow and cause problems, such as improperly processed and canned meat, sausage, vegetables, and some digestive systems of infants.*

The cases of botulism poisoning in infants was never definitely traced beyond a shadow of a doubt to honey because almost all foods can have botulism spores. The infants did have one thing in common, they were all constipated. This allowed the Clostridium spores they had eaten the opportunity to grow in the infant's warm, moist, no oxygen intestine. As infants grow older, they develop their own resistance to small levels of botulism, but as new infants they do not have this resistance. It's important to understand that anything an infant is fed except mother's milk right from the source could be the problem in infants.

The heat required to kill the Clostridium botulinum spores themselves would render the honey unpalatable. No producer/packers I am aware of do anything other than raise the temperature of their honey to 100-150ºF to retard crystallization.

If the risk of eating honey because of botulism problems were that well documented, the FDA would require a warning label on honey. They don't because not enough specific data were generated.

This situation is of concern to us all, but with the proper explanation to restauranteurs I think most fear can be removed.

HONEY GRANULATION

Question - The honey we buy in our stores never sugars. We don't like honey that has sugared or granulated even after it is melted. Please tell me how to prevent the honey from our bees from sugaring or granulating.

Answer - *Most of the honey you see in the store has been crystallized, but it has been liquefied by heating to a high temperature to retard further crystallization, and much of it is filtered or strained very finely.*

Melting honey should not have any detrimental effect on it, unless you are heating to a high temperature (over 140º) and holding it at this temperature too long. It is the combination of temperature and time that deteriorates honey. If you are going to heat it to a high temperature, then it should be only for a minute before it is cooled.

Some honeys which are high in dextrose, are almost impossible to hold in the liquid state. We would suggest that you heat, strain, and bottle in clean containers while it is still hot. You should store it at a fairly high temperature, about 75-80º.

Don't pass up the chance to "cream" some of your honey. It is granulated honey, but the crystals are so small that the texture is as smooth as butter. You can obtain finely granulated honey by purchasing a small container of creamed honey at the store and then thoroughly mixing it with a couple gallons of your honey. Store it in a cool basement or refrigerator until all the honey is finely granulated. I bet you will like the results!

WHO COULD RESIST THIS GIFT PACK?

CANOLA AS A NECTAR SOURCE

Question - I would like to find out about the Canola plant. What kind of honey production can be harvested from the plant and how many hives per acre should one use?

Bud Webb
Frederick, Maryland

Answer - Good question since the popularity and supply of Canola Oil is growing every year.

First we have to thank Canadian plant researchers and breeders for recognizing and developing a commercially practical plant with this relative of the mustard family.

Canola first came into commercial popularity in the early 1970's in Canada and Europe. The acreage devoted to this oil seed crop has grown ever since. In the last few years American farmers have shown interest in growing Canola, which could be great for American beekeepers.

Canola is an excellent early nectar producer. Depending on all the variables of rain, temperature, wind, etc., that can affect honey production, it is not unusual for a beekeeper to produce 100 lbs. plus of honey per hive per acre. The Canola farmer also benefits from a 20 to 50 percent increase in seed set.

To every good side sometimes there is a down side. The down side in the case of Canola is that honey produced from Canola granulates incredibly quickly. Within a matter of time as short as one week from when honey is capped within the hive, it will start to granulate. Once granulation has begun, it proceeds at a very rapid pace, making extraction virtually impossible. Beekeepers will have to closely monitor honey storage from hives on Canola and remove supers or individual frames as soon as 90% are capped and then extract immediately.

On the positive side, a fast-granulating honey like Canola make excellent creamed honey when care is used to insure fine granulation. For an early dependable source of nectar, pollen and honey, Canola can only help American beekeepers.

Question - I would like to ask you a question that I haven't been able to get answered from the local beekeepers. My friends and I have a few colonies of bees from which we collect the honey just for our own use. I would guess most of the nectar source in this area is white clover, as there is a lot of pasture land used for grazing. It seems like the heaviest honey flow is from late June through July. We pull and extract our supers in September. This year for the first time in over ten years of beekeeping, we found a lot of bright red honey. I have never found this before, nor have the other beekeepers I have talked with. I first noticed the red honey in the combs that are broken when you pry the supers apart. At the time, I didn't pay close enough attention to where in the hive they were situated to determine when it was possibly collected.

Something else I wonder about is that there is a strawberry grower within a two mile radius of the hives. I know that at the end of the strawberry season, there are a lot of overripe berries going to waste. I don't know if the bees would gather anything from them, or if they did, whether it would be colored. Also, the hives are within half a mile of town, and I wonder if their source is from something there. But both of these facts have been the same for years for those hives. Another fact is that there are some fields in the area planted for wildflower seed production (mostly coneflowers, however).

Bob Hassick
Pampa, Texas

Answer - *There are several answers that come to mind regarding your red honey mystery.*

First of all, your theory that the bees could be collecting strawberry juice from the overripe or broken strawberries is quite reasonable and could very well be the answer. You could readily decide this by tasting the nectar I would think. It should at least have a hint of strawberry smell or flavor to it if this were the case. Honey bees are opportunist insects and if they find a source of fruit juice such as broken strawberries or broken grapes, they will readily collect that juice and store it in combs as nectar. However, they will not break the fruits, themselves, first. This must be done by other insects or by the natural process of the fruit becoming overripe and splitting open.

Another possibility is that your bees are collecting nectar from sumac trees. This tree, which is common in the middle eastern U.S., produces an amber red colored honey when large concentrations of sumac are available. However, I have never heard of a bright red honey as you describe. You might check with long-time beekeepers in your area to find out if their bees have ever collected red nectar before and, if so, what the tree of plant source is.

Still another possibility is that you may have a candy or juice factory in your area or perhaps even a homeowner who had a red syrupy material setting outside. Since honey bees are opportunists, they may have been attracted to the red syrup and readily collected it over a period of time until the source was exhausted. We have heard of this happening before also.

One last possibility is that honey was placed over red pollen in the combs. Some trees or plants produce red pollen (examples: <u>Lamium purpureum</u> and <u>Aesculus carnea</u>) and if clear or amber colored honey were placed over the red pollen in combs, it would give the honey a reddish tinge. This is more of a long-shot answer, but I thought I should mention it since the possibility does exist.

Question - I have a question about the honey (about 350 pounds) removed from five hives in the same yard last August. The yard it located on the edge of a small oak and shagbark hickory woods. All the frames were drawn out and capped.

When the frames were uncapped for extracting, the honey was unusually light amber in color and extremely thin - almost like water. It is stored in 50 pounds plastic buckets. It shows no signs of fermentation and is slow to crystallize.

Research indicates that it might be honeydew. However, it was stated that the honeydew was to use a polarized light. Honey would deflect the polarized light to the left while honeydew would deflect the light to the right.

I do not have access to a polarized light and I probably wouldn't know how to use it if I did. Is there a simple, easily performed test method for differentiating between honey and honeydew? I could not find another test recommendation.

If what I have is honeydew, is there a use for it? Should it be fed back to the bees? Can it be blended with other honey?

Orville Kersey
Crownpoint, IN

Answer - *Your guess as to the origin of your "honey" is as good as any since you are in the area and would have observed the bees on the leaves of trees in the vicinity of your colonies gathering the honeydew.*

Honeydew gathered from the excretions of aphids feeding on plants is highly sought after in parts of northern Europe and is generally much darker than most honeys. *Honeydew is more likely to be gathered during times of nectar dearth. It may be found on the leaves of certain trees, in general, oak, beech, tulip poplar, ash, elm, hickory, maple, poplar, linden, some fruit trees and some evergreens such as fir, cedar and spruce.*

Your research is correct on using polarized light to determine honey from honeydew. Here's why. To quote from page 894 of The Hive and the Honey Bee, *"It has long been known that natural honey is levorotatory (ie. rotates the plane to the left). This is largely due to the excess of levulose (so called because it is levorotatory) over dextrose (dextro, or right-rotatory). Honeydew generally shows dextrorotation to some degree due, in part, to differing dextrose and levulose content, but more to the presence of the characteristic sugars of honeydew, melezitose and erlose, which are strongly dextrorotatory."*

Other comparisons of honey and honeydew show honeydew to be higher in pH, higher sugars, acidity, ash and nitrogen. Using sophisticated analysis techniques, the above differences can also be used to distinguish between honey and honeydew.

To make a long story short, it doesn't appear that there is any easy method to tell one from the other. Taste and color are often, but not always, good giveaways. But if the honeydew/honey is light in color and isn't granulating and has an agreeable flavor, you may have a marketing advantage if you care to sell this unique product. If not, keep it for yourself, blend it with other honeys or feed it back to the bees this spring. However, don't use honeydew as an over winter feed because it has a high dextrin content which can cause bee dysentery.

CHAPTER 10

COMB HONEY

COMB HONEY

Question - I started working with bees in March of this year. My father, who had several hives when he was a boy, is helping me. We bought a shallow super which holds 28 split boxes. The boxes have the measurements of 4 1/4 x 4 1/4 x 1 7/8 inches, and the wax is 17 x 4 1/8 inches. We put it together, and put it on top of the hive with a queen excluder between the hive body and the super about the first of June.

A few weeks later we looked at the super and nothing had happened. They hadn't even gone up into it. So we put the super between the floorboard and the hive body with the queen excluder between the hive body and the super. A few weeks later we looked at the super and they hadn't built any comb in the super. We put the super on top of the hive and took off the queen excluder and poured some honey on the super and they didn't do anything. I would like to know how to get them to work on the super.

Answer - There are several ways to encourage bees to work in comb honey supers. We recommend that bees be allowed to become established in a bulk comb honey super first since the bees will start using it more readily. Then additional supers (sections) could be added between the well-worked bulk super and the body. Bait sections, sections which are partly drawn, may also help to draw the bees into the supers. Always put the section super above the brood nest, but never place it over a queen excluder. It is seldom that a queen will lay in a section super.

**ONE OF THE MOST BEAUTIFUL SIGHTS —
NICE COMB HONEY**

PARTIALLY FILLED SECTIONS

Question - I guess I put on too many section supers. I did very well, but I have quite a few partly filled sections. Some of these are partly capped. Would you please tell me how I can feed them back to the bees for winter feed?

Answer - The easiest way to feed comb honey back to the bees is to place the sections in a super under the brood chamber, or above the inner cover with the hole partially open so that the bees will take the honey out and store it in the brood nest. Many beekeepers who have only a few sections, uncap and extract the honey. Generally, section supers do not make good wintering space and I would not recommend their use over winter.

Question - I have set up a hive with 2 boxes for the brood, one deep for honey, and on top of that I have a shallow with Ross Rounds set up. This is a nice strong hive and they are doing a great job. The deep is almost all capped over and ready for extracting. My problem is they haven't even touched the comb section above. It's like they don't want to bother with it. Should I put the Ross Round section right above the brood section? How can I get them to work it?

Mark Gosswiller
Boise, ID

Answer - You've run into one of the classic problems of producing fine comb honey. You are correct in your observations that honey bees do not like "our" little boxes or circles or half comb devices that we demand that they fill with comb honey. It's unnatural for them to expend the energy to work with these small disjointed segments. They like bigger expanses that have easy winter access or use later as brood area.

The key to producing excellent comb honey is to have a large population of bees, a consistent nectar source and then restrict the bees to where they can store this nectar. In other words, you can't give them a choice of supers or give them several supers. Put one or two on at a time, depending on your view of the nectar available and force them to use them as their only storage area. They will then fill them up.

While crowding is the key to successful comb honey production, at the same time the beekeeper must walk a fine line between producing comb honey instead of producing swarms! Many beekeepers have told us that double queening has solved the problem for them. Why not try it next season?

FRAME SPACING FOR CUT COMB HONEY PRODUCTION

Question - What frame spacing do you recommend for ten-frame supers (6 5/8") for cut comb honey production?

Answer - Standard spacing for frames is 1 3/8" and this is the spacing that should be used while the bees are drawing foundation into combs. If wider spacing is used dur-

ing this drawing of the combs, you're apt to end up with crooked combs and bridge and brace comb. *After the combs have been drawn using the regular spacing, you're free to space them wider. In a ten-frame super for the production of cut comb honey you might space nine frames evenly across the super.*

COMB HONEY

Question - I started beekeeping 2 years ago and the first year had one hive. I was interested in getting comb honey. I had 2 standard 10-frame hives with a box of empty combs of 28 sections. It took all summer but they filled 20 full sections. I was perfectly satisfied. In the early spring, I bought more queens and bees and started them in a 2 story 10-frame hives. Fed them good and everything was first class material. Not one comb did I get in the fall. So I tried putting the sections between the 10-frame so that the bees would have to go through the sections to get to the upper hive, but still they wouldn't make combs in the sections. Each hive had plenty of bees, so many that they were on the outside just crawling around. Why weren't those bees out getting nectar? I didn't get any comb honey the second fall either but the bees did gather enough honey to last them through the winter. What would happen if next spring I took the top hive along with 2 or 3 pounds of bees, gave them a new queen and on top of this put on a box of sections for combs? We have wonderful sweet clover here and I can sell a lot of comb honey if I can get it.

Answer - Comb honey is difficult to produce unless conditions are just about right. Not all bees want to produce comb honey, but most of them can be crowded into the supers with a fair degree of success. The queen appears to be important, so some selection is necessary if you want colonies which will produce comb honey. Colonies which, when crowded, do not go into the comb honey supers during a honey flow should be requeened or used to produce extracted honey.

Our suggestion would be that you give the queen all the space she needs for brood rearing until the start of the main nectar flow. At that time, reduce the colony to one story which should contain the queen, brood of all ages, and a frame or two of honey. Shake most of the bees in front of this hive and add a comb honey super on top. When the bees start to work on the second super it should be placed directly on top of the brood nest and the first super placed above it. A third one may be added on top at this time. Remove the supers as soon as they are full and capped. The extra supers of brood should be stacked 3 to 4 high on a weak colony or used to make increase.

Question - You said in one answer that bees do not winter well in section comb honey supers. This worries me as I have two colonies with one section comb honey super on each one. What is the difference?

In the spring I suppose there will be brood in the supers. How will I get the brood out of them? Will they still be all right for section comb honey?

Answer - *The problem with wintering in section comb honey supers is that there are too many divisions of wood for good clustering. If bees are to cluster well, they should be able to start on a few cells that are empty and should be able to move the cluster on comb in either direction at will. You may get away with it, but it is not the best practice. We would suggest you always have on hand at least a full shallow super for wintering each colony.*

The best way to remove the bees and any brood that might be in the comb honey supers would be to place them under the hive body in the early spring, probably in early April. When the queen is again laying in the hive body, which is now on top, remove the comb honey super. It would be well to add a shallow super at this time as it will give the colony more room to grow. Large colonies are necessary for a good crop of honey, regardless of the kind.

These old sections will not be usable for a honey crop. Never use sections over except a couple of clean ones filled with comb which has not had brood reared in it. These may be placed in the super as bait sections to get the bees into the super more rapidly. These bait sections usually have a tough midrib and should not be sold.

PROPOLIS

Question - Can you please give us any information regarding propolis? Propolis has been reviewed over here (South Levon, England) on television. Also, how could it be consumed beneficially by human beings and for what purpose?

Answer - Propolis is a generic term for certain sticky substances secreted by plants, mainly trees. It occurs, for instance, on the buds of poplars and on the trunks of pines. This propolis is also used for the material the bees use in the hive, which is usually a mixture of (plant) propolis and beeswax. The word is derived from the Greek pro (before) and polis (the city) – propolis being used to make the protective shields at the entrance to a hive. (Bee Research Association).

Propolis is indeed finding more uses than a sealant, fixative, or nuisance within the beehive. Propolis is being used in both Europe and Asia in minor medical applications. Small cuts or minor burns are reported to respond well to a tincture-administered propolis.

Research on propolis has been done in the Soviet Union for some time. Scientists have used propolis to treat tuberculosis and stomach ulcers. In Scandinavia, much research has been done since the late '60s. Findings from studies conducted in the Soviet Union indicate that propolis may have some effectiveness when taken as a food supplement by individuals who have infections of the mouth, nose, or throat, Infections of the kidney, bladder, and prostate have also reportedly been relieved by propolis. Propolis may well become a benefit to modern medicine. However, to say that propolis is a cure-all or a new and powerful antibiotic would not only be premature but irresponsible. Nevertheless, propolis is attracting increasing interest and research.

ELIMINATION OF PROPOLIS

Question - As the elimination of propolis is desirable in modern beekeeping, I coated 16 hive bodies and frames with hot paraffin except where comb fastens to the frame. This took about 15 minutes per unit and is not expensive. What is your opinion on this?

Answer - It will be interesting to see how your experiment comes out where you coated hot paraffin on the inside of the hive and also the frame. We would guess that the bees would still continue their propolizing and burr combing and that after a few months such colonies would have just as much propolis and just as much burr combing as their nature and inheritance required in spite of the fact that you had coated the wooden material with the hot paraffin. There will be a residual effect from the paraffin for a short period. This is partly because of the fact that the odor of paraffin is slightly repulsive to honey bees and they will tend to stay away from the paraffin coated wood until such time as the odor has been eliminated. It would be our guess, however, that eventually they will go ahead and use the propolis and the burr combing.

**PROPOLIS BEING USED TO FILL IN
BETWEEN FRAMES AND SUPERS**

PROPOLIZING THE QUEEN EXCLUDER

Question - Why did the bees close up the openings in the queen excluder that was placed between the brood chamber and the first super? The openings were plugged with a thick wax. This was noticed when I started to put on the second super. There was no activity at all in the first super.

Answer - One possibility is that the thick wax which you mention the bees have used to close the openings in the queen excluder may be propolis rather than wax. Propolis is a vegetable waxlike, sticky material which the bees use to close down unwanted openings in their hive. Many times they will almost completely close the main entrance to the hive with this material. When it is warm the propolis may be quite sticky. When the weather is cold, the propolis becomes quite hard and brittle.

I would also guess that the honeyflow has not been too great in your area. Sometimes when the honeyflow is rather meager the bees will tend to store all of their surplus honey in the bottom unit and fail to go up into the supers. This is especially true when the frames in the super contain foundation rather than built combs.

We would suggest that you remove the queen excluder and then put the super back on. This will make it easier for the bees to move up into the super. Then, if there is enough of a honeyflow to warrant it, the bees will start to work in the super, building combs and putting the surplus honey in them.

CHAPTER 12

STINGS/VENOM

Question - The use of the "bee sting therapy" in the treatment of such auto-immune diseases as arthritis is not new. (*The Hive and the Honey Bee* by Dadant suggests the idea is probably centuries old.) I have questions on more recent speculations that bee venom may have therapeutic potential for use in other diseases. These speculations are based partly on the observations that:

1) The immune system is stimulated to produce "quantities" of white blood cells immediately upon injection of bee venom into the body; and,

2) The white blood cells will engage and destroy ANY invaders that happen to be in their vicinity, even though they were originally produced by the immune system in response to the bee venom.

There is also more general speculation that any such regular "exercising" of the immune system may translate into a "strengthening" of the system.

The questions, then, are: Is there any hard data at all to suggest that beekeepers are not getting some diseases? What do we really know about what "bee sting therapy" can (and cannot) do?

Robert Carlson
W. Babylon, NY

Answer - You are not alone in your interest in the bee venom therapy. There have been some remarkable documented cures and major improvements in people with various debilitating diseases. There also have been cases where another person with the same disease taking the same bee venom therapy showed no change one way or another. There have been enough "cures" though to draw more attention to bee venom therapy from the public and the medical community.

My opinion (An opinion is like a nose, everyone has one) is that bee venom therapy may be contributing to some remarkable positive health benefits in some people; enough so that we should support medical research into its use.

The real problem is that the giant pharmaceutical companies don't do any research unless they are positive that they can develop a totally new drug which they alone can patent. You can't patent bee venom, so why bother?

There is an organization which is championing the gathering of all information available on bee venom therapy. This is the American Apitherapy Society, Inc. at P.O. Box 54, Hartland Four Corners, VT 05049, Ph. 1-800-823-3460. I would encourage you to contact them and support their efforts.

Question - I enjoy working with bees, but when I am stung by them, I experience pain and swelling at the site of the sting. This can last for days and once when stung near my ankle, I could hardly walk for about a week. Is this normal?

Craig Foy
Bakersfield, CA

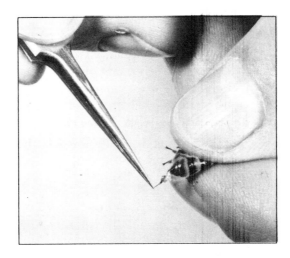

Answer - *I'm glad your getting stung occasionally has not taken away your joy of working with your bees. Anyone who says that getting stung does not hurt, is a better man than I am. It does hurt and even now after many years I sometimes get a little puffiness after a sting.*

As strange as it may sound, what you are experiencing is normal for the most part. Sensitivity to stings can change from no reaction to large local reaction and back again at anytime. This changing of reaction is tied to aging, state of health or unhealth, medication one may be taking, etc. As a generality, though, the more one gets stung, the more "immune", if you will, one becomes and less pain and swelling appears. Sometimes some people do not ever become desensitized to natural stings. They then must go to an allergy specialist who will desensitize them over a longer period of time using bee venom. Here again the procedure is not 100% effective 100% of the time and additional treatments may be necessary.

What you are experiencing, no matter how uncomfortable it is, is a natural local reaction by your body to this foreign substance, venom. It is not life threatening. Perhaps receiving fewer stings is just a matter of putting on more protective clothing or using better gloves. Keep enjoying your bees.

*See the following chart I borrowed from the 1992 edition of **The Hive and the Honey Bee.***

Table 3. Normal and allergic reactions to insect stings.
From Chapter 27, Allergy to Venomous Insects by Justin O. Schmidt.
1) Normal, non-allergic reactions at the time of the sting: Pain, sometimes sharp and piercing; Burning, or itching burn; Redness (erythema) around the sting site; A white area (wheal) immediately surrounding the sting puncture mark; Swelling (edema); Tenderness to touch.
2) Normal, non-allergic reaction hours or days after the sting: Itching;

Residual redness; A small brown or red damage spot at the puncture site; Swelling at the sting site.

3) *Large local reactions: Massive swelling (angioedema) around the sting site extending over an area of 10 cm or more and frequently increasing in size for 24 to 72 hours. Sometimes lasting up to a week in duration.*

4) *Cutaneous allergic reactions: Urticaria (hives, nettle rash) anywhere on the skin; Angioedema (massive swelling) remote from the sting site; Generalized pruritis (itching) of the skin; Generalized erythema (redness) of the skin remote from the sting site.*

5) *Non life-threatening systemic allergic reactions: allergic rhinitis or conjunctivitis; Minor respiratory symptoms; Abdominal cramps; Severe gastrointestinal upset; Weakness; Fear or other subjective feelings.*

6) *Life-threatening systemic allergic reactions: Shock; Unconsciousness; Hypotension or fainting; Respiratory distress (difficulty in breathing); Laryngeal blockage (massive swelling in the throat).*

Question - I was wondering if you could tell me how much venom is in the average working bee?

Mary Burnett
Raton, NM

Answer - *We did some digging on the amount of bee venom in the average working bee. The literature points out that a small amount of venom is already present in bees two days old. However, the amount builds up until about 18 days when production is no longer continued. There was quite a bit of variation in the literature on the amount of venom in a single bee. Dr. Hoenig of the University of Minnesota had a review years ago. In this one, he mentioned that the venom of the adult bee would be around 0.3 mg. Then a device for collecting venom from bees was developed. Using large quantities of bees, they were able to collect only about 0.1 mg of venom per bee. Any way you look at it, it's not a very large quantity. Incidentally, they found that the venom they collected from the adult bees ran 35 to 50 per cent solids.*

Question - How does the potency of bee sting compare with that of other stinging insects - hornets, bumble bees, yellow jackets, wasps, etc.?

Fred Fulton
Montgomery, AL

Answer - *You asked about the potency of honey bee stings as compared to other stinging insects. If you have ever been stung by honey bees, hornets, wasps, bumble bees, etc., you would probably say that they all were potent. But, you will probably also say they all were also different. The venoms of these insects and most others consist of proteins, peptides and miscellaneous substances. Most venoms have in common the ability to cause pain and destroy red blood cells. Stinging insects, while* they have similar venom components, are still all different in the quantity of these components. *From an evolutionary standpoint, this is not surprising. There are only a few ways to discourage hive/nest attackers and that is pain, tissue damage, swelling, death, etc., using venom. As a result, evolution selection has caused similar venom to be developed, but not exact copies in different species. I'll leave the potency question open for your judgment.*

STINGS

Question - First let me state that I am a novice in beekeeping. I have just started and have a lot to learn. I am a school teacher. The other day (in October) I took a group of students out to watch while I robbed a colony. Several of the students were stung. One boy was stung near the left eye. The next day his eye was swollen shut. During the noon hour I gave each student a small spoonful of comb honey. Within thirty minutes his eye was opening and two hours later it was open. In your opinion, did the honey have anything to do with this or was it likely that his system just cared for it? I am very interested in finding out about this.

Answer - I was unable to find any reference to the use of a spoonful of comb honey as a treatment following a reaction to a sting. I would like to suggest that such demonstrations to a group of children should be done only when the conditions are optimum for having gentle bees in the hive. Any given hive will respond differently to varying conditions, weather as well as the honeyflow conditions, sometimes from day to day.

Before they are taken to see the bees, the children should all be told how very important it is to remove the stinger immediately by scraping the stinger out rather than picking it out with two fingers. Squeezing the sting out with two fingers is like squeezing the bulb on the end of an eye dropper – it forces the poison from the poison sacs into the flesh. However, if the stinger is scraped off, even though a small portion of the stinger should remain in the flesh, the poison sacs will be removed with only a minimum of the poison ever getting into the flesh.

Such things as hot compresses, or baking soda, dampened and placed on the point of where the sting entered may help, probably from a psychological viewpoint. However, there is one product called Sting-Kill which will neutralize the effects of the acid if the material is placed on the flesh as quickly as possible after having been stung. Sting-Kill is available from Dadant & Sons, Inc. and it might be wise to have a supply on hand for future field-trips.

Question - My 16 month old son wandered to a beehive with a stick and ended up with 20-25 stings. I took his clothes off and applied vinegar with a scraping motion to try and remove the stingers. Then I applied soda water. To my surprise he did not swell at all. He had a high fever the next morning and was very tired. I have been told it would be dangerous for him to get a number of stings again. Is this true? In this case, if the bees had continued to sting him would he have gone into shock?

Answer - We don't know why your son did not swell as a result of the stings, except that he is apparently not allergic to bee venom. His reaction seems normal and your treatment was not wasted in that it probably removed many of the stings with a minimum of poison being injected into his system. There seems to be some doubt as to the value of external medication but you should check with your physician to get a treatment for bee stings. Usually an antihistamine is effective.

EFFECTS OF STINGS

Question - I have six colonies of bees now; two headed by Dadant hybrid queens, two by 3 banded Italian and two by Golden Italian. A few weeks ago, while I was working with them, two or three stung me. In a few minutes I became hot all over and began to itch and burn all over and again yesterday as I was working with one of the Goldens, two bees stung me, one on the right arm above the elbow, the other on the right little finger. I paid no attention to the stings but went ahead working with them. But again in a few minutes I became hot and began to itch and burn all over from the bottoms of my feet to the top of my head.

Up until these two occasions, I had been stung many times with no ill effects at all. Now I don't want to get rid of my bees if I don't have to. So what can I do? Is there a medicine I can take before going to work with them than will prevent this burning and itching and make it safe for me to work with the bees.

Of course I only keep bees as a hobby and although it's only a hobby, I would still like to be able to handle bees that would put up at least a little surplus. I have had a few of the yellow Italian bees that were real gentle but the ones I have now are about like the three-banded. I have never tried the Caucasians. What do you recommend? Should I get rid of my bees? Could I requeen with a more gentle bee say in September?

Answer - Our suggestion is that you see your doctor and follow his advice. It won't be safe for you to keep bees if you are becoming more sensitive to their stings. As a rule, people who start to become allergic to bee stings get progressively worse as time goes on. Many doctors will prescribe some drug, usually an antihistamine. There is also a series of shots which will desensitize some people to the effects of bee stings. These may be of help to you and your doctor would be able to administer the shots.

If there is a most gentle bee, it is the Midnite, a hybrid bee bred from Caucasian lines. Of course, even this bee has stings so your best course is to see your doctor and investigate your reactions to bee stings further.

CHAPTER 13

DISEASES AND TREATMENTS

Question - I'm curious as to what causes a strong, sour odor to suddenly come from a hive. Do you have an explanation? The hive appears to be strong and active and has produced a surplus of honey this year.

Harold Green
Huntsville, AL

Answer - The first thing that comes to mind is the brood disease European foulbrood. The odor of the decaying larvae infected with European Foulbrood varies, but typically it can be described as sour. Your description of the hive as strong and active and having produced a surplus of honey doesn't describe a hive so infected with EFB (European Foulbrood) or American foulbrood that one can smell it strongly.

My guess is that maybe you are smelling the nectar ripening from one of the fall flowers that bees work this time of year. I have smelled the nectar ripening from the flowers bees have visited this time of year and generally it is a strong pungent odor that one can smell many feet from the hive.

You need to open your hive and inspect the brood area for a diagnosis of EFB or AFB. There is a good description of these diseases in The Hive and the Honey Bee. If it is EFB, it is treatable with Terramycin. If it isn't EFB, you may be in store for a unique fall honey that you may want to leave on the hive for the colony's winter stores.

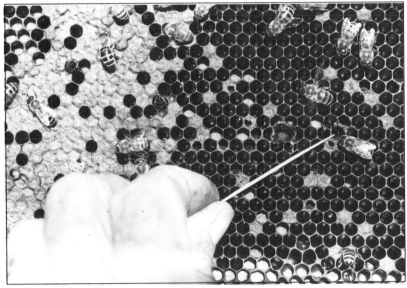

DEAD, SUNKEN PUPAE FROM EITHER AFB OR EFB

Question - Can I feed Fumidil-B in honey to my bees?

Jack Budewitz
Wisconsin

Answer - *I've never heard anyone do it, but there is no reason that you couldn't, provided the following precautions are kept in mind:*
• Obviously, be as sure as you can that the honey came from disease-free colonies so you do not introduce or spread AFB or EFB, in particular.
• You may have to dilute the honey somewhat in order to get the Fumidil to dissolve in a thinner solution.
• You must also be very careful that the Fumidil-B and honey mixture is not mistakenly used by you for human consumption or that it is not fed when surplus supers are on hives that you intend to use for extracting.

Question - What is the recommended schedule for treating a newly installed package (or swarm) of bees with Terramycin, Apistan Strip, Fumidil-B and menthol?

Fred C. Hardin
Lampasas, TX

Answer - *Check your package bee supplier to see what treatments or medications these bees have already been exposed to. Most good suppliers will have treated for all diseases and parasites already. If they don't they will be out of business sooner or later. Assuming the package bees have been treated and you installed them in new equipment and new foundation, they will be fine until late summer or early fall when you will want to treat for mites, for sure, and other diseases as needed.*

Question - Regarding the article "Strictly for the Hobbyist" in the September 1992 issue of the *American Bee Journal*, the illustration on page 578 alludes to the use of extender patties as a carrier for Terramycin.
What would be the formula for the patty for both tracheal mites and foulbrood control?
What size patty is used for each hive?

Dennis Ivy
Spendora, TX

Answer - *As you picked up, extender patties have been studied for use in both foulbrood and tracheal mite control. Let me state up front that extender patties "seem" to work when properly prepared, but no agency has amassed enough information to allow a formal registration of the control method for foulbrood or tracheal mites.*
When making an extender patty for the control of foulbrood using Terramycin, remember that it is available in several formulations with each formulation having its own specific instructions. Please read the primary or secondary label instructions to be sure that the Terramycin you are looking at is registered for use on

honey bee colonies. If the wrong formulation of Terramycin is used, it may kill your colony or contaminate any surplus honey on your hive.

For one colony 1/2 cup vegetable shortening mixed with one cup of table sugar and the appropriate amount of Terramycin for treatment of one colony is mixed together, formed into a patty and placed on the top bars over the brood nest.

There is some evidence to suggest that vegetable oil that may get on young bees interferes with tracheal mites locating and parasitizing the young bees.

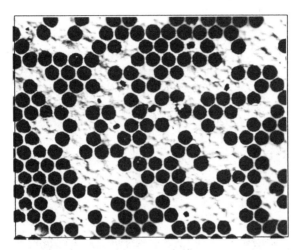

A SIGN OF FOULBROOD IS BROOD CAPPINGS WITH PUNCTURES AND HOLES

Question - I have two colonies. For the winter, on one of the hives, above the two deep supers, I left one shallow super of honey. On the other hive, above its two deep supers, I left two shallow supers of honey. This is my question: when I treat the hives in the spring, with Terramycin and Fumidil-B, will the medication linger in the supers? I am concerned that when I harvest the honey, I'll have contamination from the early treatments.

Joe Peczynski
Livonia, MI

Answer - If you treat a minimum of three to four weeks before what you feel is your major nectar flow, the chances of the bees incorporating any of this medication into the nectar coming in to make honey is slim and none. The bees will have had enough time, generally, to consume the medication and clean up any nectar medication residue produced. Remember, that in early spring and all through spring that the bees are eating stored honey, pollen and any dribbles of nectar coming in as fast as they can in order to provide the food and energy to raise thousands of new workers.

That's not to say that if you over medicate or medicate just before the major nectar flow that you could not contaminate the honey stored. Always follow label directions on all medications and chemicals and you will produce a healthy product.

If you are still concerned, remove the supers before medicating in the spring. This would only be practical if the cluster has not already moved into the supers in late winter or early spring.

Question - I put Terramycin and Crisco mix patties in all my hives. I put the patties on the hives in December. The first week of February I opened the hives to see more than half of the patties still on top of the frames.

Do I need to clean this out before I put the supers back on for honey production? We have had a mild winter, so the bees have left the super full.

Bob Wilson
Abilene, TX

Answer - The point that concerns me is that you mentioned your Terramycin Crisco® mixture, but didn't mention sugar as an ingredient. Honey bees do not like Terramycin or Crisco®, that's why we add large amounts of sugar in the mixture to coerce the bees into at least trying to consume this greasy concoction. Bees will try to "eat" just about anything if it is sweet enough. If you did not put sugar in your patty mixture, then the bees have been taking pieces of it outside to dispose of it as hive debris. If you did add sugar and the winter was as mild as you say and the bees had plenty of food, they probably would ignore it also because it doesn't taste as good and is harder to use than honey or nectar.

You do need to take it out of the hive before the honey flow because of possible contamination of the honey with Terramycin.

Sounds like your bees are strong and you are ready.

Question - In regard to the recent article on Fuller candy in the (*ABJ* November '93), I wondered, would it be possible to add Terramycin or Fumidil-B to the candy if it were added as the candy has cooled somewhat? Or, if not, how could the bees be treated not feeding syrup? Please respond. Thanks.
P.S. I have been helped many times by the Q & A's in your column.

John Warner
Ponca City, OK

Answer - I appreciate you saying that my Q & A column has helped you in the past. Hopefully, the answer to this question will also help.

In order to get either Terramycin and/or Fumidil-B mixed properly into anything, it has to be in a fairly liquid state. When that Fuller candy mixture is in a final liquid state, it is extremely hot. This heat would inactivate the Terramycin and-or

Fumidil-B. Unfortunately, the only way that I know of to effectively feed Fumidil-B in a way that it actually benefits the bees is to feed it in a liquid. Terramycin can be fed in a liquid or in a powdered sugar mix.

As you mentioned in your letter, if these antibiotics could be fed in a candy it would be helpful, but they are just too heat sensitive.

Question - I was hurt in 1984 and while I couldn't take care of my bees all the hives got foulbrood and died. Last year I started rendering all the old combs and burning the frames. I had a lot of Pierce plastic frames and I scraped them down to the midrib. I soaked them and then washed them under pressure to loosen and wash away cocoon from the midrib. Now my question is, what can I soak them in to kill the foulbrood spores? – A solution of Purex or Clorox? How strong? Anything better to sterilize the plastic combs? I want to soak them for the time it takes to kill the AFB spores and then spray them with clean water to take away the smell.

Lloyd Krekau
Cottage Grove, OR

Answer - If you have removed all wax and larvae skins etc., from your plastic frames using a pressure washer, I'm sure you have removed most AFB spores that may have been present. Most AFB spores, if present, would have been trapped in the wax comb and larval skins. With these completely removed, there will not be any reservoir of spores to cause re-infection. Even if you soaked your frames in pure bleach, it wouldn't penetrate the wax to kill any spores and might ruin the combs. We suggest you contact the manufacturer to ask for his sterilization suggestions. Some plastic combs and foundation can be boiled to remove disease spores. Remember that the hive bodies and supers you have used must also be cleaned completely before installing your newly cleaned frames. Use a blow torch to scorch the insides of the boxes.

Best of luck to you as you restart your beekeeping operation. I'm sure you will do well.

Question - Is it possible to get American Foulbrood from foundations made of wax that could have been contaminated from an infected hive?

Danny Washburn
Monroe, UT

Answer - I suppose anything is possible, but this is highly unlikely coming from the one or two largest manufacturers. I would be more nervous if the foundation was being made by someone in his garage or barn who didn't have the very efficient wax melting and filtering systems. Even if the heating of the wax to a liquid form didn't render the AFB spores inviable, the final plate filtering of the wax should remove many small particles in the wax. These filtering systems are so effective that a yellow wax can be made virtually white by simply filtering.

Question - I have learned that the dramatic losses of honey bees over the last few years especially the spring of 1996, was caused not by tracheal or varroa mites, but by a suspected virus that these mites were carrying and giving to honey bees. I have never heard what the name of the virus is, if it's a new virus or old virus gone bad. All I hear is the term parasite mite syndrome (PMS) to explain this away. What's the real story?

Nick Vogele
Hornell, NY

Answer - I don't know if there is an end to the story as yet. What I do know is that there have been recent large bee losses in Europe, where they have had Varroa and Tracheal mites much longer than we, that are not related to just mites. There seems to be a synergistic action taking place among Acute Paralysis Virus (APV) and Slow Paralysis Virus (SPV) and Varroa mites. These viruses have been identified for years in honey bees, but have never been shown to be the cause of death in honey bees in nature prior to the arrival of Varroa Mites.

Here is a quote found in Bee World 77 (3): 117-119 1996 from Dr. Brenda Ball IACR Rothamsted, England. Dr. Ball is a world authority on bee viruses.

"Studies on pathogen incidence in (varroa) infected colonies on the European mainland and, more recently in England, have shown that acute paralysis virus (APV) and slow paralysis virus (SPV), respectively, are primary causes of both adult bee and brood mortality. These findings were unexpected as previous observations over many years, before the arrival of V. Jacobsoni, had failed to detect either APV or SPV as a cause of mortality in nature. Without this knowledge, the true significance of the role of the mite as an activator and vector of inapparent virus infections would not have been recognized."

Sometimes, even with the best goals and intentions in mind, science still takes many years to unravel complicated biological interactions. This is probably the case here.

Question - My bees were destroyed totally last Fall by American Foulbrood. All advice I get locally - destroy everything, boxes, frames and combs.

I have in front of me a Russian manual for beekeepers where it instructs how to disinfect combs. Don't we have anything in this country of a chemical nature for disinfection? Please advise! I need help badly. Thanks.

Zygmunt Bugaj
San Diego, CA

Answer - I am sorry to hear of your experience with American Foulbrood. Let me tell you what I know about the disease. As you probably know, the active stage of American foulbrood is caused by a bacterium called Bacillus larvae. The bacterium multiplies in the gut of the honey bee larva and the larva ultimately dies from the spreading infection. Bacillus larvae has a very efficient and effective means of surviving when conditions become inhospitable. It is called a spore. These inactive spores wait for conditions to improve so that they may grow and reproduce. In the meantime, these spores are impervious to many chemical disinfectants and heat that can be used in the hive. To quote from The Hive and the honey Bee"...spores can remain viable indefinitely on beekeeping equipment."

So why doesn't AFB kill every colony that it comes in contact with? Different honey bee colonies show different degrees of natural resistance to the disease. The methods of resistance are these: removal of diseased larva, larval resistance, filtering by the proventricular valve in the honey bee and bacterial inhibitors added to brood food by nurse bees. In fact, there are some queen breeders who have bred and are advertising for sale AFB resistant stock.

Since AFB is everywhere in every colony, the advice to burn your hives completely while theoretically sound is in my opinion extreme. I have seen colonies that always had 5 or 6 infected larvae during the year. Does this mean they have AFB? Yes, it does but, it also means they were controlling it. However, before you make a decision check with your state bee inspector to find out what the laws require in your state.

If the state law does not require that the equipment be burned, my suggestion to you would be to scrape and clean your equipment of all wax, propolis and other extraneous material. Burn your old combs and start with new ones. Scorch the insides of your hive bodies, covers and bottoms with a torch. Then, seek AFB resistant stock and when you re-establish your colonies again, start a treatment schedule using Terramycin which will reduce the active stage of the infection. My guess is that your colonies will be able to handle AFB better and also cope with mites (with your help) and other diseases found in and on honey bees.

Question - I have been managing colonies for a large commercial queen breeder in the Tropics for some time. These colonies are in single deep boxes with a medium super over an excluder. They are used as a supply for bulk bees to support queen mating nucs and queen banks. At times they are stressed heavily, although we feed many supplements.

This season several other beekeepers and I noticed a strange phenomenon in one yard. There were several colonies with blackened deposits on the ground in front of the hive, almost painted in appearance. One hive, however, was most puzzling. For a period of two months a pile of bees developed (6-7 bees thick), on the ground in front of the hive. There were many dead and dying bees that were trembling on the pile. There did not seem to be any weakened colonies and nothing suspicious was observed in any of the colonies. No pesticides or insecticides were used in the yard. We worked the yard approximately every two weeks. The elevation is approximately 1500 feet and often cloudy. Is my suspicion of a virus correct? Possibly Chronic Paralysis Virus?

Kevin Fick
Kealakekua, HI

Answer - You must have done some research in the literature to suspect a virus as the agent in your bee death question. Eliminating herbicides, pesticides and poisonous plants as a cause leaves just a virus as the possible culprit. Without isolating a virus from your diseased bees in a laboratory setting, one may never be completely certain. But, with symptoms you describe, I would have to agree with you that Chronic Bee Paralysis Virus (CBPV) sure looks like a likely candidate.

Working in the commercial queen breeding industry is interesting because this disease seems, from the literature, to be hereditary. Here's a quote from page 1110 in The Hive and the Honey Bee: "How paralysis spreads from bee to bee and why only isolated colonies in an apiary are decimated by the disease are unknown. The fact that the virus does multiply in pupae suggests that hereditable factors influencing susceptibility are operative. Attempts to document this aspect were successful when Rinderer et al. (1975), and Kulincevic and Rothenbuhler (1975) demonstrated that they could select strains that were more susceptible to a hairless black syndrome that proved to be CBPV (Rinderer and Green, 1976)."

Readers do you have any ideas as to the cause?

Question - I was a hobby beekeeper for 25 years in Romania and now for two years in San Bernadino, CA. Just in this short time I have lost 3 colonies with chalkbrood disease. The brood becomes dry in the cells with a white-gray color.

I have given the treatments of Terra patties, Terramycin and Apistan two times a year.

I want to know what causes this disease. What kind of treatment can I apply for chalkbrood? What preventative treatment can I use to avoid it in the future?

John Poptelecan
Fontana, CA

Answer - Chalkbrood is a fungal disease that infects larvae three to four days after egg hatch. The infected larvae are soon overgrown with white cotton-like threads called mycelia. The larva is killed as this fungus mycelia feeds on it. Finally, a hard white mass forms which is referred to as a chalkbrood mummy. There are no chemicals or antibiotics commercially available for the control of chalkbrood. There are few things which can help though:

1) A - Chalkbrood infections are more noticed during cool wet, damp periods of the year, spring being most common. When the weather becomes warmer and dryer, the majority of the time chalkbrood lessens or disappears entirely.

B - Do your utmost to place your hive in an area that has good air drainage, is not continuously cool and damp and do not wrap, pack or otherwise restrict hive ventilation. The bees will ventilate the hive well if the beekeeper does not interfere with the process.

2) Cull the old dark combs. These old combs are a reservoir of all kinds of contamination and infection. Replace old, dark brood combs with new foundation every 3-4 years.

3) A - Many times the bee's housecleaning ability is not as good as other strains or races of bees. Bees that exhibit the good housecleaning trait will identify and remove dead, dying and diseased brood quickly. This helps reduce reinfection.

B - Replace the queen.

4) I've heard that one tablespoon of standard household bleach in 1 gallon of sugar syrup fed during a non-honey flow period helps control chalkbrood. This is not an approved or authorized treatment, but may be very safe to use. Some essential oils are also said to be anti-fungal in nature.

5) Keep very healthy colonies that are not weakened from mites, diseases or other maladies. A large strong healthy colony can effectively deal with many problems.

Don't be discouraged. You'll be on top of the problem quickly and things will turn around for you.

**CHALK BROOD MUMMIES ON A
BOTTOM BOARD**

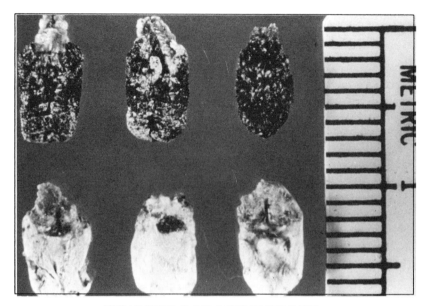

**CHALKBROOD MUMMIES UP
MAGNIFIED**

CHAPTER 14

PARASITIC MITES & TREATMENTS

Question - I read the Q & A section of the magazine regarding the use of Apistan strips. I am a little concerned as my Apistan strips have been in year round in the brood chambers and it was said that prolonged use will contaminate the combs and the honey and that eventually the mites will become immune to Apistan. I would like to keep Apistan only after the honey flow and the beginning of the honey flow. Mistakenly, I left the Apistan in all spring and summer (when I should have removed them) and now winter is coming and I am afraid they have been in too long. Would it be harmful to leave them in until next spring or should I remove them during the winter? I am off schedule, so what should I do? I still have supers with honey in my hives which will be removed soon.

How is the honey contaminated - is it harmful to eat? So far the honey appears to be in excellent condition and abundant.

My other question is, how and when is the best time to move colonies 400-500 feet (between the old location and new location there are shrubs, trees and a small shed)?

Drago Nizetich
Lakewood, CA

Answer - Everyone can make a mistake and not remove Apistan strips from all colonies. You are correct in stating that leaving the Apistan strips in longer than label directions may cause problems.

When the strips are left in past the label instructions time period, this is what happens. Within the first week to ten days approximately 97% of the adult mites on your bees die. This leaves 3% who for one reason or another are immune or resistant to some degree. Additional mites become exposed and die to the active ingredient (fluvalinate) as parasitized young bees emerge from their cells up to the 42 days when strips would be removed. What you have left is the 3% of mites who survived and can now breed. Genetics being what they are, the offspring of these survivors will exhibit various degrees of resistance. Some like mom and dad mite will be resistant totally, others will have no resistance to Apistan. Without Apistan in the hive, these breed until the next treatment time when because of all the crossbreeding between resistant and on resistant mites, the resistance gene is diluted. You treat again and kill 96-97% of adults again.

If you leave the strips in, the mites are exposed to sub-lethal amounts of fluvalinate over long periods of time. This means that you are forcing the selection of fluvalinate "super" mites. The only mites breeding will be the ones who are surviving the constant exposure to fluvalinate. Pretty soon no amount of fluvalinate will kill them without also killing the bees.

My suggestion is to remove your strips and plan on applying "new" strips in the spring.

Apistan strips are made by combining the fluvalinate with a plastic and form-

ing it into a strip. On the surface the bees brush against the fluvalinate, picking it up on their bodies and the mites' bodies. The bees brush against each other spreading it throughout the hive. As you can see, it gets everywhere. I do not believe that excessive residues have been found in honey by beekeepers using the Apistan strips, according to label directions. Be assured that it is in your honey, wax, pollen–everywhere. Whether it is in concentrations high enough to be illegal or dangerous for consumers only an analysis of the honey would tell.

In moving your colonies, you may have to make two moves–one about 2 miles away to get them far enough away so they won't return to the old location for a week or so. Then, your final move to your new location. They won't be able to return to the last location because of distance and will stay where they are. Seems like a lot of work, but if you just move them the 500 feet, all the foragers will go back to the old empty location and you might lose them.

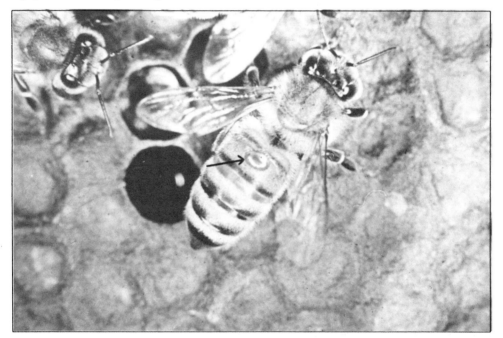

VARROA MITE ON WORKER BEE

Reader Comment - Recently I read a letter in the *Classroom* where the reader was asking about a natural extract-based miticide. The answer given was that we shouldn't try it because it hasn't been in bee hives long enough.

Am I missing something, or isn't 6000 years long enough? As long as bees and plants have coexisted on this earth, the oil of that plant has been brought back to hives in pitch for propolis, nectar for honey, and pollen. I started using a natural miticide three years ago and I credit it for saving my bees.

Doug Hamilton
St. Helens, OR

Answer - Dear Mr. Hamilton:

Thanks for your comment directed towards my answer to a question in the June 1995 Classroom. By the tone of your comment, you had a knee jerk reaction to my answer about the safety of using a "natural miticide". If you had calmly read my whole answer and thought about it, I believe your response would be more logical.

*To recap for those just joining in, a beekeeper wrote in about some product that is purported to be a combination of plant extracts, vegetable oil and or petroleum jelly that supposedly kills varroa mites and what my opinion was on this. My response was that **a chemical, is a chemical,** whether it comes from a plant or a laboratory–that because of the unique nature of the inside of a hive (with workers, queens, drones, eggs, larvae, pupae, honey, pollen, wax etc. all open and exposed to whatever is in or added to the hive unit) that we should be careful. Proper and complete testing should be done to establish the efficacy and safety for bees and humans of any product used in this environment.*

*What's the deal with thinking that plant extracts are somehow less lethal and more friendly killers than any other chemicals? It's a rough world out there in nature. There is constant competition for the resources of air, food, water and space. Plants and animals either kill, are killed, parasitize, develop survival means or leave to find these resources. Some plants are not eaten by bugs, birds or mammals because the "natural poison" contained in these plants will kill whatever eats them. Either way, natural or synthesized, **death is death.***

Since you did not read carefully, the very first paragraph to my answer, here it is again for you. Hope it helps.

There are probably thousands of compounds from plants, molds and fungi that separately or in combination with other ingredients will kill varroa and tracheal mites. There are probably only hundreds of these compounds that will kill these mites and not generally be lethal to honey bees. But on the road to finding these chemicals, we have to be very sure of what, where, and how they will be used, to be used safely."

Editor's Note – Also, to be in compliance with the law in the United States, this product must be okayed by the Environmental Protection Agency.

Question - I understand that there are some plant extracts or essential oils that will kill varroa mites, but which are harmless to the bees themselves. They are apparently mixed with vegetable oils or petroleum jelly?

James Tipton
Glade Park, CO

Answer - There are probably thousands of compounds from plants, molds, and fungi that separately or in combination with other ingredients will kill varroa and tracheal mite. There are probably only hundreds of these compounds that will kill these mites and not generally be lethal to honey bees. But on the road to finding these chemicals we have to be very sure of what, where and how they will be used, to be used safely.

Anytime we think about killing a bug (varroa or tracheal mites) on a bug (honey bee), we have to consider this unique undertaking and the complex environment it takes place in.

First let's talk a little about killing a mite. Let's use an analogy. Humans are not hosts to any mammalian parasites that I am aware of, but let's pretend they are. Let's pretend there are parasitic creatures the size of mice which attach themselves to us and that we cannot normally remove ourselves. How would a chemical agent be found to kill this parasitic mouse and not harm the person? It would be tough to do - kill a mammal on a mammal. Any chemical that would kill these parasitic mice would probably at least make us sick, miss work, may affect our reproductive systems and if we were weak from the parasite already, it might kill us. The same thing happens with the honey bees and mites. Using currently approved chemicals of a natural origin, hive organization is disrupted, some adult honey bees die, the queens don't lay as much and some but not all of the mites are killed – and this is with a tested product.

Because of their life cycle, we have to kill these mites in the hive. Let's look at this environment: Open cells of food-pollen, honey, nectar, larval food–all exposed to the hive atmosphere and also potentially in contact with any honey bee and what she is carrying on her body.

Beeswax, because of its chemical composition, is a chemical sponge that absorbs any fat soluble chemicals, natural or man made. Honey bees then attempt to raise their young on this beeswax, feed them food that at some time was stored in this beeswax and also fed to the nurse bees caring for them.

I didn't mean to be overly windy in response to your question, but I do want us to realize that this is a complex question with complex outcomes. Just because something is labeled as plant extracts or plant essential oils, don't believe for one minute that these "natural" chemicals won't kill as readily as something man-made. They will! And without extensive testing and field trials, how do we know if this stuff kills 100% of mites and no adult bees or does it kill 100% of mites, no adult bees and 50% of the larvae or does it kill 10% of the mites and is absorbed into the beeswax so that the queen won't lay in the cells. We could obviously go on and on with different scenarios.

We don't know the answers to all of these questions about new potential mite compounds because the time and scientific effort has not been devoted to finding out. I trust our University and USDA labs enough that they could find out if given the opportunity. Support them with the funds necessary and they could tell you the most effective way to live with mites and produce a honey crop.

My last and shortest advice is: DO NOT USE ANY CHEMICALS IN, ON OR AROUND A HIVE OF HONEY BEES WHICH HAVE NOT BEEN APPROVED BY OUR GOVERNMENT FOR USE IN BEE HIVES.

Question - I have some questions that have been bugging me for quite some while.

Are Varroa and Tracheal Mites attracted to other insects? Most research that I am aware of deals with treating our hives which I agree we must all do, but I was wondering if there is something that can be done to at least reduce the number of mites in a given area such as host insects or bait traps that might be attractive to mites?

Mike Kazak
Charleston, NH

Answer - Varroa has been found regularly on wasps, hornets and bumble bees. An interesting note is that many beekeepers in England feel that the recent introduction of Varroa mites came from bumble bees. Why? Because in Europe and England many vegetable crops are grown under acres and acres of greenhouses and these greenhouses find it more convenient to use bumble bees as pollinators and bumble bees were imported to England from the Netherlands for this purpose. Whether this is, in fact, true or not may never be proven, but it is entirely possible.

I haven't heard that tracheal mites parasitize any other insects, but logic says that any insect visiting a flower is fair game for tracheal mites.

I think that trying to control these mites on all managed and feral populations would be virtually impossible but, a good question for researchers.

Question - I lost about 40 percent of my colonies this past winter from a combination of tracheal mite infestation and late winter starvation.

Because of my mite problem, can I install new package bees on the old equipment?

B. Ross
Flagstaff, AZ

Answer - Yes, any mites that had killed the individual bees would have also sealed their own fate by killing their host. Any mites that may have been outside of the bee tracheae will have died in a few days without a honey bee to feed on.

Remember, you will have to treat this fall with an acaricide before your bees go into winter. Tracheal mites are everywhere.

Question - We put Apistan into our hives this past winter according to package instructions. We inserted 2 strips per hive Oct. 18 and removed them Nov. 27, 1992. We went to check our hives last week and found honey in some of our hives. Our question is: Is this honey consumable? If not, how should we go about getting rid of it.

John & Deloris Davis
Lake Elsinore, CA

Answer - *Apistan, when used as directed, is a very good control for Varroa mites only. If the honey was stored while the strips were on, you might want to let the bees consume this to be on the safe side. If the honey has been stored after the removal of the strips, it will be safe for human consumption.*

As a side note, we are getting more reports of the tracheal mite causing honey bee mortality. Be sure that you are treating with a product designed to kill tracheal mites, too.

Question - I was wondering if you know the kind of bug that is killed by tobacco smoke. A beekeeper from Yugoslavia has mentioned a bug maybe Asian that is killed this way and I was wondering if you had ever heard of one.

Virl Dowdy
Phoenix, AZ

Answer - *Your beekeeper friend from Yugoslavia was obviously referring to the external honey bee mite called Varroa jacobsoni which has infested bees over much of the world in the last 30 years.*

When tobacco is added to a bee smoker the nicotine from that smoke will kill a certain percentage of varroa mites in a colony of bees. Nicotine, as you know, is a poison and is quite toxic to many insects.

Unfortunately, the tobacco smoke treatment is not effective enough to kill most of the mites, so those that remain quickly reproduce and reinfest the colony.

Apistan, which is produced by Zoecon Corporation in the United States, as well as much of the rest of the world, is much more effective in controlling varroa mites.

Question - I live in town and I have six hives of bees in my back yard. I have kept from two to eight hives for years, and I have not seen this happen.

The first hot day, over 75º F the bees in two of my hives came out and covered the front sides and the ground in front. They seemed to lick at each other, as if they might have honey or sugar water on them. This was not consistent, however.

They kept this up only on hot days, for about 6 weeks. I lost one hive, the other one I removed the frames and put in two frames of honey and two of brood. The queen seemed O.K. and they are doing good. The other hive had no queen and only a handful of bees. Q. - What do you think was wrong? Q. - Since bees seem to like soda water, could the artificial sweeteners be the problem? Q. - Can honey go bad in the hive? Q. - How can you tell if you have bee mites?

Roy Goodman
Hillsboro, TX

Answer - *Your first question about your crawling bees in hot weather is a tough one. I really don't know exactly from your description, but you say they were inconsistent in this behavior. So, let's look at the possibilities.*

Crawling behavior such as this can be a sign of heavy tracheal mite infestation, heavy nosema infection, pesticide poisoning and some other less common problems.

Whatever it was, it seems to have led to the death of one of your hives. The other you gave a boost of brood and honey which may have helped break the cycle of whatever was happening.

With all the information I have obtained about colony loss from tracheal mites with not exactly the same, but similar reports, I'll stick my neck out and say that's what it was. Tracheal mite populations fluctuate during the year. In spring when there is a nectar source or the beekeeper is feeding his colony, the honey bee population increases faster than the tracheal mite can breed and the percentage of infestation drops. This may or may not have been what caused the one colony to apparently recover.

Your second question dealt with honey bees having access to artificial sweeteners. I don't know about this question. The honey bee's digestive system would have to process a large quantity of foreign chemicals, i.e. the artificial sweetener and they wouldn't benefit from it. It couldn't be good for them, but I also couldn't tell you if it caused the above symptoms.

And your next-to-the-last question asked about honey going bad in the hive. Honey may at times, because of excess moisture, start fermenting in the hive at which time it becomes useless for the bees. This may cause dysentery and characteristic "spotting" of your hives. Sometimes honey also will granulate in the comb. This is a natural process which will not hurt the honey or the bees, however.

Question - How do I know if I need an acaricide?

Answer - Bee mites can be so destructive that they can weaken or kill an entire colony. Unless your hive has been inspected by someone trained to diagnose both Varroa and tracheal mites, you cannot be certain that your hives are mite-free. It is safest to treat with a pesticide specifically formulated for mites (an acaricide). .

Question - I want to know, how long it takes for the mites to kill a hive of bees?
Charles Damron
Wyandotte, MI

Answer - Let's assume that your colony before infestation with mites was in perfect health, no AFB, EFB, NOSEMA, no pesticide exposure, no over-wintering food shortage, plenty of pollen, a young active queen and on and on. If your bees got either Varroa or Tracheal mites separately, then it would take three to five years before complete loss. If you have both mites and any or all of the previously listed problems, it could be much quicker that you experience colony death. And it doesn't matter what season they die either. But higher losses are noticed in spring after trying to over winter.

There isn't much time to stall in treating your colonies with Apistan, or men-thol because both mites have been here several years and have spread virtually everywhere. Good Luck!

VARROA MITES ON DRONE LARVA

Comments: I am referring to your article in the *American Bee Journal* September 1992, page 580, in which you vigorously object to the use of formic acid for mite control.

May I suggest that you refer to the article in the May 1991 article in the same magazine page 311 by T.P. Liu of Fairview College in Alberta Canada. Said article states that formic acid has become the most favored acaricide in Europe and is in use in several other countries.

As a chemical engineer, I do not regard formic acid as being any more toxic to humans that many of the insecticides which the local farmers spread by the ton. It is quite similar to acetic acid which is commonly used in photo developing, and which, in dilute solution, is known as vinegar. Being quite volatile, its longevity in honey and wax would be quite minimal.

Our government is slow in using information from aboard. Other nations have had this mite problem and have been working on it for some years. Can you think of any reason why we should not profit by their experience?

John Young
Rhinebeck, N.Y.

Answer - *As a trained chemical engineer, you have a better understanding, I'm sure, of formic acid and its properties which should also help you realize the inherent dangers involved.*

First, let me tell you what I know and if I'm wrong, please correct me.

(1) The use of formic acid is illegal for use on colonies of honey bees.

(2) Being illegal means that if it is used and that one consumer in a million has an adverse reaction to it because of high levels in the honey that he or she just consumed, you're going to have a lawsuit that will choke a cow. If the news media gets ahold of the story, then honey's image is reduced because of fear of its use by consumers.

(3) I too have heard of beekeepers experimenting with formic acid illegally. The formic acid available on the market is in its concentrated form. You as a chemical engineer know what this means: Respirators, gloves, goggles and the knowledge and ability to remember how to dilute this dangerous and caustic material. Do I add the water to the formic acid or the formic acid to the water? If you do it the wrong way, it blows up. Stop me if I'm wrong, John.

(4) I've talked to several beekeepers who have experimented with formic acid and all have stopped using it because of chemical burns on their hands and the problem with breathing vapors.

(5) If you talk to European beekeepers like I have, they tell you that they have to treat every six weeks or so with formic acid. This controls the mites, but the queens have to be replaced every year because whereas the workers are only exposed to the formic acid vapors once or twice before they die, the queen gets a full shot every time. After a year of this, she quits laying and is superseded or is replaced by the beekeeper.

So, with the information I have formic acid in the form available in the U.S. is illegal to use, dangerous to use and results in queen replacement every year.

And this is supposed to help beekeepers? I agree with you that we should learn from other nations' experiences. That means we should look at the positives and the negatives.

Thanks for getting my adrenaline going.

Question - I have a question I hope you can answer. I'm wondering about hot oil mist I see advertised as a way of getting rid of Varroa and Tracheal parasites. Is there any documented proof that it really works? Has anyone used it and then wrote about it?

Virl Dowdy
Phoenix, AZ

Answer - *Thank you for your recent question on hot oil mist treatment for mite control.*

There is no documented proof of consistent control in formal research done in the U.S. or Canada.

There has been some testing done primarily in the old Soviet Union on this method, but they got equally good results from just raising the temperature of the bees' treatment environment without using oil.

I would hold off on any investment in hot oil equipment until you can gather a more complete picture of the process, procedure and results. Ask the seller for names of customers and contact them to see what they thought of the product.

Question - I am a hobby beekeeper with five hives. One of my hives is obviously dying even though it is too cold to open up for a close examination. What I see happening is that the bees are tossing out dead bees by the hundreds. One day it warmed up enough for a quick look inside. As this hive is a first-year hive, I didn't harvest any honey. The hive has lots and lots of honey. To make a long story short, I've narrowed my suspicions to two. Either something happened to the queen during the season and all the bees left are old and dying or I'm having problems with tracheal mites. What's bothering me the most is that I'm only guessing at the mite problem. What should I do?

David Guffin
Roy, WA

Answer - Your "problems" could be all of the above or none of the above or a combination thereof. Even though there have been hundreds of research and scientific papers written on and about honey bees, because of the nature of this super organism, the honey bee colony, diagnosis of this type of problem is equal parts art and science.

You could very well have a tracheal mite problem, but as you suggest, without a microscope and some training, it is just a guess. I would contact your state apiarist, Mr. James Bach at 206-586-5306 and he should be able to help you determine if tracheal mites are a problem or not.

Another possibility is that because of long confinement, the dead and dying just piled up and then warm weather gave the live bees a chance to clean them out. Nosema and varroa could also be problems in your hives.

In any event, your reasoning on your problem is excellent. Now you just need the means to validate it. That's where Mr. Bach comes in. Give him a call; he'll be happy to help.

Question - I live in the West Texas Panhandle where there are very few bees. In fact I don't know of any other beekeepers in my country. I am concerned about whether I should treat my bees for Varroa mites.

Al Kirby
Bovina, TX

Answer - If this was a perfect world, your local beekeeping inspector should come out check your hives, take sample of your bees and test them for both varroa and tracheal mites. He would then report back to you on the outcome of these tests and make a recommendation.

The above scenario probably isn't going to happen. After having seen and heard of the devastation caused by varroa by beekeepers who didn't think that their bees had them, I would suggest that you treat with Apistan. For a few dollars you may protect your whole hive. I think it is a wise treatment.

Question - We all know about the honey bee's resistance to varroa mites: earlier brood development, grooming, biting and mite removal from the hive. How are honey bees resistant to tracheal mites? I'm sure these mites are too small to be groomed by the bees.

Dan Mihalyfi
Watsonville, CA

Answer - Why do some honey bees appear resistant to tracheal mites? Good question. The answer is nobody really knows. Surprising isn't it that so many claims about tracheal mite resistant bees have been made, yet the mode of bee mite resistance or control isn't known.

There are some hypotheses as to the how and why of infestation and, of course, why some honey bees are maybe resistant. Here are the three reasoned guesses:

1) The breathing tubes (tracheae) of the honey bees have openings called spiracles to the outside environment. One thought is that these spiracles are smaller in some bees, thus restricting the entrance of tracheal mites looking to set up housekeeping in that bee's tracheae.

2) The hairs surrounding the opening of the spiracle on resistant strain bees are more numerous and stiffer, thus setting up a barrier to tracheal mites.

3) Tracheal mites are attracted more to lighter colored bees than darker colored bees. So, the darker bees don't get infested as readily.

I personally like all these reasons, but as in many questions, there usually isn't just one nice clean answer. My guess is that tracheal mite resistance is a combination of these things and probably others not identified yet.

Question - When I temporarily remove Apistan strips from between frames during routine inspection, I usually find some bridge comb attached to them. I scrape it off, of course, but the wax continues to coat the surface. Is there anything I can use to remove this wax which will not also remove the fluvalinate?

Dan Hendricks
Mercer Island, WA

Answer - Remember that the fluvalinate of the Apistan strip is not just on the surface of the strip, but incorporated throughout the whole plastic matrix of the strip. As long as you do not destroy the strip or all of the fluvalinate has not already migrated to the surface, a surface cleaning (scraping) won't hurt a thing.

**APISTAN STRIPS FOR VARROA MITE
CONTROL**

Question - Is there a scent or pheromone that is very attractive to bee mites, that I could add to a sticky pad with an 8 mesh wire cover to keep the bees off, that I could put in my hive to keep the mite population down?

John Street
Kane, PA

Answer - Your question concerning if a "mite" pheromone has been identified to then use as a lure is an excellent one. Many researchers are working on just that. If they can identify a pheromone which the mites use to locate each other or a pheromone used by the mites to find a food supply, then trapping them becomes much easier. To date there have been no conclusive results along these lines. Perhaps in the near future there will be. Until then, we will have to use currently approved chemicals to control Varroa and tracheal mites.

Question - I've had a terrible time keeping Varroa mites out of my colonies. Can I leave the Apistan strips that I am using in longer than the 45 days indicated on the label directions? Is it possible to leave them in permanently?

Vettie Van Allen
Ooltewah, TN

Answer - As your letter states, the varroa mite has really made an impact on American beekeeping. Many beekeepers throughout the U.S. have lost colonies to these parasites. We are very happy that we have Apistan strips for the control of varroa. In order for this product to continue to be effective, it must be used according to label directions. These directions state that it should be used for a set period of time and then removed. The reason for this is because the longer you leave the product in the hive after the time period stated on the label, the more chance there is of the remaining mites (Apistan only kills about 97% of mites) developing resistance to the active ingredient. As you know, during your lifetime many pests have developed resistance to certain pesticides. We don't want this to happen with the varroa mite. Do not use Apistan any longer than label directions state and you will be fine and your colonies will be better. Of course, under no circumstances, should you leave Apistan strips in your hive during the honey flow period, unless you do not intend to harvest the honey for human consumption.

Question - I have several questions about mites: 1) Are mites (of any kind) seriously affecting honey production in central Utah? 2) Should small beekeepers (1 to 15 hives) routinely treat for mites? We presently do for foulbrood. 3) Is there an easy way to detect mite infestations? 4) Is there an effective method of treatment for mites?

Edward L. Jones
Monroe, UT

Answer - Addressing your first question, I didn't have a clue, but I knew somebody who did. As you probably know, Ed Bianco of the Utah State Department of Agriculture at 801-538-7184 is the head of honey bee inspection for your state. He told me that Utah does have tracheal and varroa mites, but that they have not been blamed for any serious losses in colony populations or honey production. He thought that conscientious application of approved mite control agents by beekeepers in Utah had stopped any serious problems from occurring.

All beekeepers should inspect and treat for mites. If you have mites, then it is just a matter of time before conditions allow the mite population to grow and do real damage to your colony. Once damage is seen, it is often too late. The mite population will be so large and your colony so small, that it will be hard for the bees to recover no matter what you do.

The tracheal mite is very, very small. It lives in the tracheae (breathing tubes) of the honey bee. Identification has to be done under a microscope. Mr. Bianco can tell you who offers this service in Utah. The only approved control for the tracheal mites is menthol crystals. A 50 gm bag of menthol crystals is placed on the top bars (if the daytime temperature will be between 60º and 80º F) and left for 20-25 days. The "fumes" from the menthol evaporation kill the mites in the bees' breathing tubes. Be sure to follow complete label directions when using menthol.

The varroa mite is a relatively large external parasite. It can be seen with *the naked eye on bees, especially on drone brood. When inspecting your colonies scrape off the cappings on drone brood and look for reddish brown dots about the size of a pin head. These are varroa mites. On adult bees the varroa mites look like freckles on the thoraces of the bees.*

Varroa mites are controlled using Apistan strips. The plastic strips are incorporated with a miticide which will kill 99.8% of the varroa mites when used according to label directions. Of course, the problem must be continually monitored to prevent re-infestation from surrounding feral and beekeeper colonies.

Question - Ten of my eleven colonies succumbed to varroa after an excellent honey crop in 1994. What is your advice for the 1995 season? I have treated the remaining colony with Apistan, but I have noticed that several surrounding feral colonies have died out, probably due to varroa.

Wilbur Tripp
Nauvoo, IL

Answer - Unfortunately, you have joined a growing group of U.S. beekeepers with similar stories about varroa devastation. As in most other countries where varroa has been found, the first reaction is denial that the mite can invade your personal colonies, even though you realize that colonies are being destroyed in locations or states around you.

The best avenue now is to start fresh with new mite-free bees and then begin a regular program of fall treatment with Apistan after honey removal. If you live in a heavily infested area, you may also have to treat in early spring before honey flow starts.

If you order packages from a company that uses Apistan strips and Apistan queen tabs, you'll start with colonies that are at least 99% free of varroa.

There is still the likelihood that your new colonies will become reinfested from surrounding beekeeper colonies or surviving feral colonies. Nevertheless, you will be starting with clean, healthy bees which is a big plus and since you will still have your drawn combs and hives, the packages will get off to an excellent start this spring, provided you give them plenty of sugar syrup or honey.

Question - What natural vegetable-based products have been identified that will control the varroa mite?

R. Salomon Handal
El Salvador, Central America

Answer - Here in the U.S. we are just discovering natural vegetable products that may control the varroa mite and tracheal mite. The product which seems at this early date of research to be effective is what we call vegetable shortening. This is semi-solid vegetable oil which is sold for cooking purposes. This material can be shaped into a 1/4 pound patty (disk) and put onto the top of the hive. As the honey bees try to remove this foreign substance, this vegetable "grease" gets on to the bees and disrupts the varroa mite's ability to find a suitable host.

Various aromatic oils, mostly from members of the mint family, and a newly found possibility, neem oil, have on a preliminary basis, been shown to be somewhat effective.

Question - I take many of the European bee magazines and one of the biggest advertisers the last year or so has been Biove Laboratories of France selling a product called Apivar for Varroa control in hives. They advertise it as a "new" product, but I see that the active ingredient is Amitraz. The product looks just like Miticur which was sold in the U.S. a few years ago until some beekeepers said it was killing their bees.

My Question is: If the product killed bees in the U.S., why isn't it killing honey bees in Europe? I certainly have not seen any letters from beekeepers claiming losses from Apivar.

Loxton Stevenson
Birmingham, U.K.

Answer - *Nothing like a question to challenge my political correctness, is there? As I think I've said before an opinion is like a nose, everyone's got one. Here is my opinion on your question.*

I don't think Amitrz (Miticur) killed bees in this country. As you know, Miticur was taken off the market because a few commercial beekeepers who shall remain nameless, in one section of the country said they noticed abnormal bee mortality when they used some, but not all Miticur strips for the control of Varroa mites. This all took place in 1993. Back in 1993 commercial beekeepers to save a dime were using all sorts of products off label and putting them in their hives on shop towels, wooden strips, cotton, paper towels, even spraying these chemicals directly into or onto the bees in the colonies. Some of these hives were subjected to a real chemical soup. Now it is true that when Miticur strips were put into some of these hives there was abnormal mortality. But it is even more interesting that Miticur strips out of the same package and put into hives in the same apiary did not result in higher mortality.

Anyway some entrepreneurial attorneys got involved and the makers of Miticur were sued. The manufacturer withdrew the product, settled out of court and vowed to never again re-enter the beekeeping market in the U.S. because of the way the industry turned on them when they had spent hundreds of thousands of dollars in bringing this begged-for product to market.

Let's fast forward through the years 1993, 1994, 1995, 1996 and the winter of 1997. Beekeepers in the U.S. lost tens of thousands of colonies and they were not using an Amitraz product. They were using the other approved Varroa strip. How come no one sued the maker of this product's socks off? Sounds fishy to me.

The Amitraz product in not killing bees in Europe because it alone did not kill bees in the U.S. Remember, this is my opinion.

Question - As a hobby beekeeper, what are my chances for breeding varroa resistant hives? I do not plan to medicate or purchase queens, but just breed from the survivors.

William Dieffenbach
Los Altos, CA

Answer - *What you are practicing now is survival of the fittest. Darwin spoke of this as he elucidated what he saw in action in remote natural locations. And this could certainly happen in your situation, but it is highly unlikely for the simple reason that you, using this method, have little control over mating. Without getting into the unique genetics of mating drones, who only have half as many chromosomes as the virgin queen, there are some more basic problems.*

In a natural mating situation the virgin queen will fly up to 10 to 15 miles away from her colony and then mate with up to 10 to 12 drones who may have come from as many colonies. Some or none of these colonies may exhibit the grooming behavior necessary to remove varroa from members of the colony. Then, are these positive traits as a dominant genetic code going to be passed on from the drone to the eggs produced by the queen? The odds are not great for this happening. Now remember, this scenario has to happen with each drone and the sperm he donates.

For arguments sake, let's say that in a colony on July 15th, there are 50,000 workers. Of this amount, 100 have by chance inherited a gene that makes them exceptional groomers. They will bite and nip the varroa off of workers they encounter. This is great except that there are not enough of these super groomers to stop the varroa from continued breeding and multiplying. And, in 6 to 8 weeks these super groomers are dead. Will more super groomers develop by chance? Will there be more or fewer? And, then, say you want to raise queens from this queen. Which drone will be contributing his genes? The problem is that in a natural setting it's not like taking one cow into the field to mate with a bull.

Researchers all over the world, using a variety of control mating schemes and artificial insemination, are trying to accomplish the same thing you are doing.

Because of this knowledge and technique, they are trying to bypass the shortcomings of a natural uncontrolled honey bee mating situation. They will probably reach the goal before anyone else.

Question - I would much prefer not to use chemicals in the hive. I have had for the past several years limited success with the Crisco® patty and Buckfast queens for tracheal mite control.

This past spring I purchased 30 hives with the new mite resistant Yugoslavian strain. This turned out to be an experience in itself. Most of the queens were extremely small, 7 were drone layers and the other 23 were superseded within 60 days – all of this happening during one spring honey flow. Lesson learned, buyer beware, know your queen breeder.

Would you please detail the pros and cons of heat treating for mite control, (120º and corn oil). I have been unable to find any information about it except from the manufacturer.

Bill Weinhoffer
Sparta, IL

Answer - Sorry about your problem with the Yugo queen breeder. Unfortunately, it does happen. Hope you were able to get some remuneration from him or her.

My knowledge of the pros and cons of heat treating for mite control is like most everyone else's here in the U.S. - limited. All I can do is recite back to you what is in the literature and what some of our researchers have told me.

First, let's realize that we are trying to kill a bug on a bug. Tough to do without killing or damaging the good bug, no matter what technique you are using.

Killing tracheal mites by using heat has not had as much attention as killing the varroa mite. Tracheal mites can be killed with heat while in the honey bee, but it is a somewhat time-consuming and difficult job to do properly. The process requires the beekeeper to remove every bee from a hive, putting them in a screened box. This box of bees is then heated, in some form of closed "oven" but with an oxygen supply to 117º for 15 min. This will kill most tracheal and varroa mites. If your temperature is too high or maintained too long, you kill both mites and bees. Too little and you kill fewer mites. Remember, you never kill every mite. So, a possible breeding population is left. Also, reinfestation from surrounding feral and managed colonies occurs rapidly. From what I've read, control is no greater than the use of some chemicals, but obviously no chemical residues are left. On the down side, unless we have only a few colonies, the labor involved in heat treatment for mite control makes this procedure impractical.

Question - When I started beekeeping, I followed instructions and prepared the Crisco® extender patties by placing a gob between two sheets of wax paper and rolling it out about 3/16" thick. Then, years later, when I heard that the vegetable oil helps rid the hive of tracheal mites, I wondered if the surface area available for evaporation around the edge between the two sheets of paper was adequate. Nothing I have read about tests on the mites made mention of whether the grease was covered so or not. Then, recently I read advice to smear the grease on the top bars with a hive tool. Settle this for me, please. For effectiveness against the mites, is it okay for the patty to be covered with wax paper or should the patty be fully exposed?

Dan Hendricks
Mercer Island, WA

Answer - I'm not sure anyone really knows how much surface area is or is not necessary for optimum positive effect of vegetable oil patties if in fact, it is having an effect at all.

As you know, the logic behind using vegetable oil patties (or just vegetable oil) is that one of the volatile oils in it gives off a "smell" similar to the smell of the newly emerged bees exoskeleton cuticle as it hardens. When a bee chews its way out of its cell, its outer covering is somewhat soft and pliable and takes sometime to "airdry", if you will, and harden. The thought by researchers is that, if the vegetable oil patty is giving off a smell similar to the smell of a newly emerged bee, that mites who use this smell to home in on a new host will get confused and not find the proper-aged host.

There has been quite a bit of variability of results in formal tests on how many new bees get infected when a vegetable oil patty is or is not used. Sometimes it seems to work great; other times mite infestation is just as high as newly emerged bees, but death from them is actually less.

The newer line of reasoning and research is that vegetable oil may be disrupting mites seeking new homes, but that the antibiotic Terramycin generally incorporated into these patties may be having a more positive effect.

It seems that when Terramycin is used in these patties, that bees can still have tracheal mites, but the mites don't bother them as badly. This information would suggest that either the mites are introducing an organism into the bees as the mites feed, or the open wounds are letting an organism in that the Terramycin is controlling.

At this time, it looks as though Terramycin may be more important than the patties alone.

Question - I'm a small-scale beekeeper, having (been one for) 20 some years. I'm sure you hear from a lot of hobbyists about new and earth-shattering "discoveries". However, I have a discovery of sorts that has so far worked very well for me in controlling tracheal mites in my bees. The problem I had using menthol was that it wouldn't vaporize until the temperature reached 60º F or more. In early spring, say March in my area, maybe even April and early May some years, I would set and watch my bees dwindle because of tracheal mites. Nothing could be done because it wasn't warm enough to vaporize the menthol. Then I thought of adding alcohol to the menthol. The alcohol was the vehicle to vaporize the menthol. I've used rubbing alcohol, whiskey and even panol liniment with good success. The panol liniment worked especially well as it had lots of vapory stuff in it along with clove oil. I don't know if any one has tried this yet, but it has worked well for me. I have not lost any bees to the tracheal mites since I've done this.

Randy Stieg
Reed City, MI

Answer - I like people who "experiment". You've done a good job in thinking through a problem and you took a small risk. Dr. John Harbo, at the UADA/ARS Lab. in Baton Rouge has been working on heat and how it can be used to control mites in and on living honey bees. Heat, if high enough, can kill mites and not greatly harm the honey bees. But in his experiments he found something I thought was very interesting.

Remember the last time you exercised hard and how hot you became? Think of this as you read what Dr. Harbo found.

Bees will "fan' their wings any time there is excessive heat in the colony and when there is any kind of irritant smell that they want to get rid of. Tracheal mites found in the bee's trachea are right next to the bee's wing flight muscles. These muscles generate a lot of heat when used. If there is an irritant such as menthol or maybe in your case, clove oil or alcohol, the bees "fan" and generate a lot of heat themselves which may kill the tracheal mites. It may not be the chemical itself having an effect, but the result of internal heat produced by the bees trying to get rid of the smell.

In any event, whatever chemical vaporizes the best at low temperatures may be the one that causes the fanning response and the heat. It may be that the alcohol by itself works just as well as with the menthol. You may want to conduct your own experiment.

Keep the ideas coming.

Question - Late last fall I checked three hives of bees that I had. They had plenty of honey. This spring I went out and no bees were in the hives–no dead bees in the hive or around the hives. What do you think happened?

Lawrence Lewis
Aberdeen, ID

Answer - Sorry to hear about the loss of your colonies. Unfortunately, the observations you recounted are all too familiar. In circumstances such as yours, many bee-keepers have lost their colonies. The common factor seems to be that the bees had tracheal mites, varroa mites or both. There are two explanations for virtually all bees disappearing that I have heard. The first is that on warm days in winter, the sick and weakened bees leave the hive and do not return. When most bees are parasitized in a colony, this could explain the dramatic and quick disappearance of the bees. The second explanation is that a large quantity of the honey bee population has the genetic coding to abscond when mite or disease levels reach a certain threshold. The colony absconds at its first opportunity to leave behind the cause of the colony's problems.

Pick whichever explanation that you would like. Remember, though, that in this day and age having an adequate supply of honey on your colony is not enough. You must treat for mites if you are going to be a successful beekeeper.

SPRING DWINDLING

Question - My bees don't do well after the first of each year. Do bees get any kind of disease?

John Clifford
Oxford, N.C.

Answer - Be sure that your bees have plenty of honey stores or sugar syrup to use after the first of the year. This usually means going out and checking the colony some time during your January thaw. Add sugar syrup or honey at that time if they look to be short on stores. You can pick up one corner of the hive and determine if they are short on honey or not.

Another possibility would be disease or mites. Mites have been blamed for bee losses in your area in recent years, so we would suggest that if you continue to have colony losses or small populations, that you have your bees checked by an apiary inspector or send a sample in for inspection. The apiary inspector for your state is Karen Giroux, State Apiarist, Plant Industry Division, N.C. Dept of Agric., P.O. Box 27647, Raleigh, N.C. 27611. Tel. No. (919) 733-3610. You might also want to contact your state extension beekeeper who is Dr. John Ambrose, N.C. State Univ., P.O. Box 7626, Raleigh, N.C. 27695, Ph. No. (919) 515-3140.

Question - I received a flyer in the mail from one of the other smaller beekeeping supply businesses advertising something called formic acid. They didn't say in the advertisement what exactly it was for, but I'm assuming something for my bees. What is it?

Porter Rockwell
Carthage, IL

Answer - Your question really is a good one and I'm glad you asked about it. We in the beekeeping industry are unfortunately pawns sometimes to shysters and snake oil salesmen as anybody else.

*Formic acid is a highly dangerous and corrosive acid that in its concentrated form will burn your lungs, eyes, skin and anything else it comes in contact with. You can be severely injured or killed by using this **illegal** product. Formic acid is not approved for use in bee hives for anything.*

There have been some reports from Europe that a weak solution of formic acid and water will reduce varroa mites. The USDA is now ready to begin testing to see if there is any truth to these reports. All we need in the beekeeping industry is for residues of an unapproved concentrated chemical (like formic acid) to be found in honey or beeswax and there won't be a honey industry left. Remember a few years ago the Alar scare the apple industry endured. Apple sales went to zero for months because of all the publicity the media churned out against Alar and apples. We as an industry could not take such an occurrence. I think it is at least irresponsible and at most totally illegal for some in the industry to go for the short-term dollars and jeopardize the individual beekeepers and the industry.

TRACHEAL MITE DEAD COLONIES

Question - Please tell me what I need to do with colonies killed by tracheal mites? What should I do with honey left in the supers?

Wade Casper

Answer - *Tracheal mites need to be in contact with the breathing tubes or tracheae of the honey bee in order to be able to feed on the hemolymph (blood) of the honey bee. They can only survive for short periods of time without feeding. If your colony was killed by tracheal mites, waiting a matter of a few days will make sure that all tracheal mites are dead after the last honey bee has died. There will be no tracheal mites left to re-infect new honey bees put into the old equipment. Because of the widespread infestation of honey bees in the U.S. with tracheal mites, the chances of package bees harboring tracheal mites is very great. You will want to plan on treating your colony with menthol.*

Go ahead and leave what honey is left in the supers to help your new colony build up more quickly.

VARROA-FREE QUEENS

Question - How sure are we that queens shipped with Apistan Tabs will be free of mites and safe to put in our hives?

Lynn Ellis
Mancos, CO

Answer - *The manufacturer of Apistan tabs had as an advertisement that if the Apistan Queen Tabs are used according to directions that queens and attendants will be mite free when delivered to the customer.*

I would have to agree that an Apistan Queen Tab in a queen cage for at least three days will kill all varroa mites in that cage. Tests conducted by the USDA also conclude that 100% of the Varroa mites on bees in queen cages are killed by Apistan Queen Tabs.

Question - Considering the ever present problem of varroa mite infestation, how many times should you treat your colonies with Apistan? Last year I treated my colonies once in the fall after removing goldenrod and white aster honey and had no winter-kill. An experienced beekeeper told me to treat twice. What is your feelings on this matter?

Tim Armstrong
Ashtabula, OH

Answer - *Thanks for the question, but the more I thought about it, the harder it became to answer it in a clean black and white fashion. Here are the things (ifs) that ran through my mind:*

1) If you treat only once a year, the chance of your own population of mites building up increases because of the time between treatments. Plus you have to be aware that drifting bees will be bringing mites into a particular hive all the time.

2) If you treat too often then you are accelerating the selection, (survival of the fittest) of the mites most resistant to the mite treatment. These mites breed and, of course, you have a population of mites that is resistant to a particular miticide just as they have in Italy now.

There has to be a happy medium some place. I'm going to suggest starting off with treating in the spring before buildup and honey flows and in the fall in preparation for winter. Twice-a-year treatments are important in areas with a high density of colonies as well as high varroa mite populations. You will have to adjust your treatments as you get a better feel for the seriousness of the varroa problem in your area.

Question - In the February issue of *American Bee Journal* - page 97, Mr. Randy Stieg mentions his success with "Panol liniment' ridding the Tracheal mites from his colonies. I have been trying to find this in local drug stores and have been unsuccessful, do you have or could you contact him as to where you find it? Thanks.

Keith C. Rowe
Cable, WI

Answer - I don't want to appear overly rude, Mr. Rowe, but if you have followed my comments in the Classroom over time, I have made an effort to not promote non-registered treatments for honey bees. That is not to say that I do not appreciate the creativity and experimental nature of some beekeepers or that I am not aware of the problems that beekeepers are having. I will not promote general treatment that do not have the rigor of the registration process behind them. As I have stated before, we as an industry, both hobby and commercial, are not big enough or strong enough to have Sixty Minutes or 20/20 do a story on unapproved products being used to help produce contaminated honey. Let's stick to the rules whenever possible.

Question - I am confused about how many Apistan strips to use and how long to leave them on. I have two packages of Apistan and the directions for use on both of them are different. Can you tell me the proper way to use this stuff?

H. Carl
Sayre, PA

Answer - There has been some confusion about using Apistan because of the change in label instruction. Until the older packages of Apistan with the old instructions are used up, there may be some continuing confusion. Here are the new instructions word for word off the "new" label.

Directions for use: Use one strip for each 5 combs of bees or less in each brood chamber. (Langstroth deep frames or equivalent in other sizes). Hang the strips within two combs of the edge of the bee cluster. APISTAN strips must be in contact with brood nest bees at all times. If two deep supers are used for the brood nest, hang Apistan strips in alternate corners of the cluster, in the top and bottom super. For best chemical distribution, use APISTAN when daytime temperatures are at least 50º F.

Duration of treatment: 42 days (6 weeks) to 56 days (8 weeks).

Question - On the inside cover of the August 1995 magazine, there is an ad for Apistan. It says "Apistan doesn't contaminate honey, so you'll never face the food safety concerns of other miticides."

Does this mean you can now use Apistan during honey flow?

Jim Brown
Allyn, WA

*Answer - No, it doesn't mean that you can use it during the honey flow. What it should really say is Apistan doesn't contaminate honey, **when used according to label directions.** I think the mis-wording and the confusion it has caused was a simple big corporation left hand (marketing) not knowing what the right hand (technical) was doing. When the product, Apistan strips, is used according to label instructions, which means before or after the honey flow, no appreciable levels of fluvalinate can be found in the honey stored. Thanks for your sharp eye and the question.*

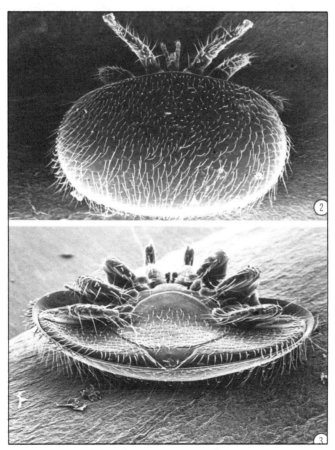

VARROA MITES UP CLOSE

Question - I have read two conflicting statement regarding Apistan–one theory says the strips continue to be effective for months after they are used because the active ingredient, fluvalinate, continues to migrate to the surface from the interior of the strip. The opposing more recent theory I have heard says that after 45 days the strips are useless since all fluvalinate on the strip has been rubbed off by the bees and no new fluvalinate migrates from the interior to the exterior of the strip. Which theory is correct?

Gerald Free
San Francisco, CA

Answer - I hope you are not setting me up on this question. It seems like one of those questions designed to have hate mail sent to me by the manufacturer and researchers because one or the other doesn't like my answer. I like to live dangerously, so here it goes.

If you put a plastic strip similar to Apistan under a high power microscope, you would see something that looks like Swiss cheese or a sponge–lots of inter-connecting spaces and channels between the plastic that give shape to the strip. In the manufacturing process these holes and spaces can be filled with special insecticides or miticides. On the surface of the strip there are some of these chemicals. If an animal, in this case a honey bee, brushes by it, some of the chemical rubs off on its body and is distributed throughout the hive. The chemical that was brushed off the surface is replaced by the chemical stored in the spaces in the strip. As this is brushed off, chemical from the interior constantly takes its place on the surface. Sooner or later all the chemical stored in all those "Swiss cheese" holes and spaces is used up and no more is available to take its place on the surface.

The way I understand it, the fluvalinate is used up after 45 days because of the above process.

PARASITIC BEE MITE CONTROL

Question - I lost 5 of my 10 hives last winter from mites. Is there any control for this? Please let me know. I will be out of the bee business if there is no control for these pests.

E. Howard Waddell
East Ridge, Tennessee

Answer - As you probably know, there are two U.S. mite parasites which are of concern to beekeepers. First, there is the tracheal mite which lives within the breathing tubes on trachea of honey bees. This mite, as you can well imagine, is very small and can only be seen by use of a microscope. The tracheal mites feed on the blood of the honey bee by piercing the wall of the trachea with very sharp mouth parts. They become so numerous in the trachea that they clog these breathing tubes in effect slowly suffocating the honey bee host.

In the U.S. only the product menthol has received a general registration which allows it to be used throughout the U.S. for the treatment of tracheal mites. Menthol must evaporate within the hive in sufficient quantities and concentrations that in the process of breathing, the menthol vapors will kill a large percentage of the mites residing within the honey bee trachea.

The other mite which is starting to cause considerable problems among bee-keepers is called the varroa mite. This mite is much larger and is an external parasite of honey bees and the developing pupae of honey bees. This mite's life cycle is somewhat different from the tracheal mite in that it damages or kills developing honey bee pupae and also feeds on the blood of adult honey bees. This parasitic mite can be seen relatively easy with the naked eye or magnifying glass.

The only product registered for use with honey bees for the control of varroa mites is called Apistan. This product is a strip which is hung within the hive and the active ingredient is spread from bee to bee and as it contacts mites on these bees, it kills them.

Question - What is the shelf life of Apistan? I cannot use all the strips at once and want to keep the unused strips for next year.

Herbert Philipsborn
Northbrook, IL

Answer - *I immediately contacted the Sandoz representative who is Oscar Coindreau. Mr. Coindreau said that the shelf life of unopened packages of Apistan is 3 years and possibly longer but for sure 3 years. His conditions were that the product should be kept out of direct sunlight and out of extreme hot temperatures.*

For opened packages, the shelf life is 2 years provided the product is kept out of extremely hot temperatures and out of direct sunlight and dry.

The active ingredient in the Apistan strip is not a vapor but a powder and, therefore, the shelf life is extremely good for the product.

Question - I enjoy the questions and answers of the *Classroom* and have one of my own: I've treated for tracheal mites using menthol the past two years and found that the bees quickly seal off the menthol using a mixture of propolis and a lot of wax. This year I timed the process and found it to be two weeks, more or less, depending on colony strength and am very concerned about the effectiveness of the treatment. (I had losses last winter even after treatment.) I overwinter in three full size hive bodies and applied the menthol the third week in August in cloth packets between the second and the upper box – the brood nest being the second extending down to the lower.

The space between the top bars on which the packets rests is fully sealed, as are the bottom bars above it. When you remove the bag, its surface is fully sealed.

I have Italians, grays and some hybrids between the two. Usually the grays use more propolis in general, but in this case all three types of bees use the same approach.

What is the experience of other beekeepers and what can you do differently, to keep this from happening?

Paul Niemeyer
Saugerties, NY

Answer - *I'm glad you enjoy the Classroom section of the ABJ and thanks for sending in your questions about the trouble you are having with menthol.*

Don't feel like the Lone Ranger, nearly every beekeeper in the country is having problems in one way or another with menthol. Menthol is one of the stop-gap products that was found early on in the struggle to control tracheal mites. When all conditions are perfect menthol works well. The problem is conditions are never perfect. Menthol, to be effective, must evaporate and be "breathed" in by the bees in order to come in contact with the tracheal mites. In sufficient concentrations it will desiccate (dry up) the mites, thereby killing them. Menthol's evaporation rate is dependent on temperature and humidity. If the temperature is too high or low, the bees do not get a proper consistent dose. Menthol is very aromatic and the bees don't like the over-powering smell in their home. It is an irritant to them. Just like when we have an overpowering odor in our home, we tie up the trash bag and take it outside. If menthol is put loose into a hive the bees pick up as many of the small pieces as possible and take it outside. When it is in a bag, they seal over it, wall it up and eliminate the problem as they see it.

The only thing you might possibly try is a spring application. Bees in late summer and early fall are preparing for winter and one thing they do is collect more propolis than usual to seal up the hive. In the spring the honey bees' goals are increasing the population of the colony and propolis collection isn't a primary goal. Maybe your menthol bags wouldn't be sealed over quite so fast. That still is no guarantee that the menthol will evaporate at the proper rate.

Question - How are mites spread between hives that are not in close proximity (1/2 mile) to other colonies?

Victor Patterson
Pearland, TX

Answer - *Mites, especially varroa, but also tracheal, have a variety of methods to expand their range and opportunities for survival.*

The most efficient method of spread for mites is the bees, themselves. Drones, particularly, will drift from hive to hive and be accepted as non-threatening in most all hives. When I say hives I should really be saying colonies, because they can drift from man-managed colonies to colonies living in a tree or side of a house or old barn back and forth all the time. Even a half mile is no real challenge because they are coming from drone congregation areas and they follow each other back to the colonies at the end of the day. Mites hitch a ride and make their way to new areas easily. Workers can also drift and bring mites, but usually not over this distance. Another survival strategy of varroa mites especially is to hop off onto a flower that a bee is visiting and then jump on the next bee visiting that flower.

Just using the above two methods of movements, you can see how mites spread so quickly across the U.S.

Question - If Varroa mites mate with their brothers and sisters in the sealed brood cell, why don't you see the results of inbreeding like you do in other animals – population falling, fatal mutations, etc.?

Trent Graham
Logan, KS

Answer - This is an excellent question! I had no idea, but started asking who in the research field could help me find an answer. I immediately called Dr. Jim Tew at OSU and he suggested a contact he had in Israel. So he got on the Web and this is what Dr. Uri Gerson in Rehevot, Israel offered. It is as interesting as the question.

Varroa breeding system: not sure whether it has ever actually been worked out, but based on literature relating to closely-related families, I can make a pretty educated guess. Likeliest it is haplodiploid, just as in bees. However, these mites have a twist of their own: female must be mated in order to produce eggs; no progeny without stimulus produced by copulation (system also call parahaploidy or pseudoarrhenotoky; common in phytoseiid mites used for biocontrol). During the subsequent cleavage the paternal component gets "lost" and is excluded from the developing embryo. Unfair, but so is life.

So, as the beekeeper asked, why do populations not collapse due to intensive inbreeding? Well, I would say there are at least four answers to that, and one open question.

1) Inbreeding is not always all that harmful. The ancient Egyptian and the relatively new Hawaiian kingdoms practiced submatings. Any weak brother or sister simply died out; Unhappy for the individual person, but keeps the royal blood clean.

2) Varroa has probably adapted to submatings long ago, which means that there had been strong pressure to remove all deleterious effects of such inbreeding.

3) More than one female often attack drone cells, so there would be opportunities to mate with non-siblings, broadening the gene pool.

4) Many cases have been reported of non-fecund Varroa; could well be due to inbreeding-induced sterility.

"And the open question: nobody really knows what happens to Varroa on its original host, Apis cerana, with its frequent absconding. We do know now that worker as well as drone cells of A. cerana are invaded by the mite, and that infested worker brood is cleaned out by the nursing bees. But why do these A. cerana-adapted mites not avoid worker cells in which they have no (evolutionary) future, and in which they (so to speak) can be said to commit suicide? Or are perhaps these mites the "inbred" ones? And, how much of its mating habits has Varroa brought over from cerana to mellifera?"

COMMENT

There have been many questions over the months about leaving Apistan strips on the hive longer than label directions or the use or reuse of strips.

From Canada, Paul Van Westendorp, manager BCMAFF Apiculture Program, recently addressed these questions in the *Beelines Newsletter.* I like it, so here it is:

Leaving the strips in for too long and reusing strips are short-sighted practices because they accelerate the development of Varroa mite resistance to Apistan. When the strips are left in for a long time, the quantity of fluvalinate (the active ingredient of the strip) that is released declines to levels too low for killing all the mites. In this "partially poisoned" environment, the few surviving mites will be the only ones that reproduce and their offspring are likely to have the same tolerance to Apistan as their parents. They, in turn, will be the only survivors reproducing another generation of resistant mites. The beekeeper unwittingly breeds for resistant mites.

After several years, we may have to apply even higher dosages of Apistan (i.e. more strips per colony, and much more expensive), until it is no longer economically viable. In desperation, producers are then also likely to resort to all kinds of "snake-oil" recipes, but in the process, may end up contaminating their bees, equipment and/or honey.

In Europe several other products are registered, including Bayverol (flumethrin). The problem with flumethrin is that it also belongs to the family of synthetic pyrethroids, as does fluvalinate. As has been observed in other mite pests, the build-up of resistance to one pyrethroid has often led to simultaneous resistance to other closely related pyrethroids. It would not be surprising if the Apistan-resistant mites of northern Italy display comparable levels of tolerance (resistance) to flumethrin.

"While there is a genuine danger to the beekeeping industry of losing Apistan as a valuable control tool, there appears no imminent arrival of other effective controls. The only alternative available in Canada is formic acid which, under favorable conditions, may be good. But formic acid, as cheap as it is, has to be handled with care and may not always be as effective as Apistan. If not applied properly, it can also be damaging to the bees. It is therefore in everyone's interest to retain Apistan's effectiveness as long as possible by: 1) using Apistan only when needed; 2) not using alternate Apistan application methods; 3) always removing strips upon treatment completion; 4) not reusing Apistan strips, and 5) alternating between Apistan and formic acid."

Question - I have 3 homemade hives; 2 hives have a double wall with 1/2 inch of insulation in between. I began in the springtime with a 3-lb. package of bees. After the last honey flow I put Apistan in. As the instructions stated, I kept it in for 45 days, but winter came early and I left them in for the winter. In early January we had a tremendous snow storm. After a few more days the temperature immediately raised into the 60's. The bees from the insulated hives started to fly 3 hours before the bees from the other hive started to fly. Because the bees were flying into the snow, I sprinkled ashes on the snow so the reflection wouldn't blind them. Should I give the hives new Apistan when spring comes? Also, do you have any advice or comments on the bees flying during the winter? Finally, what do you think of the hives with insulation?

Stanley Tabaka
La Grange, IL

Answer - *I would hold off on treating with Apistan until after your last honey flow that you will collect. When you treat next, be sure that you retrieve the strips at the prescribed time on the label directions. As stated in past Classroom sections, chances of wax contamination and hastening resistance by varroa all occur when these strips are left in or reused in colonies.*

I do have a couple comments about insulation. Insulation obviously will keep a hive warmer. When they are warmer, they eat more food, so be sure they have enough. They start rearing brood sooner, so you may have to modify your mite treatment schedule to address the mite population early growth. The bees also may fly out to defecate on days that outside may be too cold for them to return before getting chilled.

There are pluses and minuses to everything. Insulation is one of them.

Question - I know that mites are a real problem in the bee world, and I sympathize with beekeepers during these hard times. However, I am concerned that pesticides being used to treat mite infestation may find their way into the wax or propolis. Do you have any information concerning the purity of bee products obtained from treated hives?

Richard Cech
Williams, OR

Answer - *First, let's acknowledge that with today's techniques of analysis specific chemicals can be identified __easily__ down to the part per billion level. Whether a 1 ppm, ppb or ppt of a given chemical is hazardous to humans is not in my level of expertise to answer. To put parts per million (ppm) in perspective, it has been estimated that 1 ppm is equivalent to one drop of water in a 55 gallon barrel of water, while 1 part per billion (ppb) would be equivalent to one drop of water in a swimming pool full of water. I will tell you what I know and let you make your own judgment.*

Beeswax is a natural fatty acid. As a fat product it will absorb and tenaciously hold on to a wide variety of chemicals. The problem being experienced in Europe is that when they treat for mites, the beeswax is within permissible levels for one to three years. After this, the beeswax continues to absorb and retain more chemical until even the bees won't use it. The beekeepers then cut out the beeswax, melt it and sell it. Beeswax comb is not replaced yearly so anything that it is exposed to comb over its 5, 10, 20 year useable life span will be documented chemically in that beeswax.

Propolis is usually tree gums, saps and resins honey bees collect in order to coat various parts of the hive and to plug cracks and crevices. But it doesn't have to be natural sticky stuff that bees collect. It can be anything gooey and sticky – tar, paint, caulking, gum, etc. In fact, it wasn't too long ago that a buyer of propolis in the Midwest had a recall of the Chinese propolis he had purchased and sold. It had high levels of lead in it that came from a combination of lead based paint and natural lead in the environment from an industrial area where the honey bees were collecting their sticky material.

The old adage "Buyer Beware" applies here. Know your supplier.

Question - I recently received my last issue of ABJ and am quite pleased with its contents.

There was an apparent negative response to the efficacy of the use of vegetable shortening to control tracheal mites, (Don Jackson's article: "NCBA Project VIII" Pg. 564 Aug. '96).

So, what's the story, Jerry? What is the most effective treatment for tracheal mites and the time of year for treatment, if you only treat hives once a year for tracheal mites?

Tim Casey
Pall Mall, TN

Answer - *Glad you are a new subscriber to the American Bee Journal and even more glad that you are enjoying it.*

You asked what was the most effective treatment and time of treatment, if you only do it once a year, for tracheal mites.

I don't think once a year is going to cut it. Bees during spring and summer reproduce very quickly, but so do the mites. Then, when the honey bee population drops in the fall, the mites that are in dying bees aren't stupid. They leave the trachea of these bees and find a new host. Let's use this analogy: Are dogs, cats, cows, chicken, etc., just treated once a year for the fleas and mites that they are home to? No, they are not and why? Because mites build up quickly, to fill the void left when the previous ones were 90% exterminated by chemical treatments.

What's the best treatment? I don't think one has been developed yet. Remember the tracheal mite is inside the breathing tubes of the bees. It's a hard spot to get something into that will kill the mites and not harm the bees. Menthol, vegetable shortening even Apistan (when tracheal mites are outside the bees) will kill them, but timing treatments, number of treatments, and reinfestation time all have to be calculated in. Reinfestation is probably the most important. It doesn't do much good to kill 90-99% of the mites if infested workers and drones are drifting into that colony.

The short answer is treat in the fall (Sept.) and spring (April) at a minimum. Treat for other diseases, generally reduce other causes of colony stress and cross your fingers because nothing is sure!

Question - I would like to know if it is safe to put bees back into a hive that has been killed by mites. If so, how long should I wait?

Mitchell West
Decaturville, TN

Answer - Since both the Tracheal and Varroa mites require honey bees for their food supply, once the bees are gone, all the mites are generally dead within a week. It should be safe to put more bees in that colony after this time. Be sure to treat with Apistan and Menthol to keep mite populations low in bees that you will have in the future.

The exception to the above statements is when capped brood remains. This brood may eventually emerge in stored equipment. The emerging bees, whether they are drones or workers, may still have live varroa mites on them. It is fairly common to hear stories of beekeepers finding crawling varroa mites on supers of honey removed from heavily infested colonies. These supers may have been stacked in a honey house for over a week before the extracting was done. To be on the safe side, it would be wise to expose any beeless equipment with brood to freezing temperatures before installing new bees or returning it to your established colonies right away.

Question - The way Apistan strips are placed in between frames bees cannot contact the strip's total surface area as invariably the strips may lean against or on the other surface of the comb. In fact, if you do not spread the combs apart, there is less than one bee space between the medicated strip and the surface of both combs. Would this not cause the strips to be very ineffective?

Nowhere in the Apistan instructions does it say to separate the combs when inserting the strips. Would it be wise to reposition the strips after 15 to 20 days so as to expose some surfaces which may have leaned against one comb or the other?

Would it be more effective to jam the strips between combs, 90º from the recommended method, so that the side surface is in line with the cells axis?

John Bunicci
Shoreham, NY

Answer - *Thank you for your excellent question. I had often thought the same as you, but never pursued it.*

I called the people at Sandoz and was told the following. The strip was designed to fit between two frames in a ten frame hive and still allow enough space between the surface of the comb and strip to allow bee contact. That was their story and they were going to stick to it.

As beekeepers, though, we actually use the product and you are correct that in most situations there is only one side of the strip that is exposed for the bees to contact. It is obvious that there is a loss of surface area of contact. It stands to reason that if there is a loss of surface area that effectiveness is diminished. The answer is that fluvalinate is highly effective on varroa mites and that the surface area that may be lost is quickly made up for by the delivery method of fluvalinate to the surface of the strip. Fluvalinate is stored in the matrix (space between plastic molecules) of the strip. When the fluvalinate is rubbed off the surface by a passing bee, more immediately moves to the surface – the old high school chemistry lesson of material moving from a greater concentration to a lesser concentration seeking a balance or equilibrium. This method is highly efficient in the strip for moving more active material to the surface. No bee passes over this strip without picking up the fluvalinate. Since bees do pick up fluvalinate and the bees transfer fluvalinate from their bodies to their sisters' bodies and varroa is highly sensitive to fluvalinate, the whole system works in practice rather well.

You could manipulate your frames and the position of the strips, but it wouldn't really make any difference.

Question - I've noticed a lot about not being concerned about buying queens because the Apistan strip they send with her kills the varroa mites, but what about the tracheal mites?

Kay Almond
Pocatello, Idaho

Answer - *Your question dealt with queens coming through the mail and harboring varroa and tracheal mites. You are absolutely right to state that if the queen and her attendants are treated with the Apistan Queen tab that varroa are killed, but what about tracheal mites? The tracheal mites would still be there. At this time there is no treatment for tracheal mites for queen cages.*

At this stage, we can only knock down a large percentage of both mite populations within the hive and carry on with our standard hive manipulations.

Question - Are Africanized bees more resistant to either tracheal or varroa mites than more gentle types of bees?

It is well known that the White False Hellebore plant has ingredients used by vets to rid animals of fleas, lice, mites, etc. Do you suppose planting a few of these miticide around hives would help get rid of mites in the spring when the bees go for the flowers?

R. C. DeVries
Burnt Hills, NY

Answer - 1) Are Africanized bees more resistant to tracheal or Varroa Mites? This answer has two parts:

A) Yes, Africanized bees seem to tolerate Varroa mites more than honey bees of European ancestry. There does not seem to be any resistance to tracheal mites.

B) Africanized bees seem to be just as sensitive to Parasitic Mite Synchome (PMS) as European honey bees. This is why their advance seems to be stalled in the Southwest.

2) White False Hellebore: It might help, but I'm thinking the concentration of any active agent would be so slight as to have minimal effect. Give it a try. It couldn't hurt. Some beekeepers claim that various species of mint plants growing in their areas appear to inhibit heavy mite infestations. No scientific studies on this subject have been done to my knowledge, however.

Question - What is this sticky paper I keep reading about that's put on the bottom of hives?

Joseph Reed
Mount Pleasant, PA

Answer - Sticky paper is being used on the bottom of some colonies to trap hive debris to check for the presence of Varroa mites. When the mites die either naturally or with a miticide, they drop off the bee and fall to the bottom of the hive. Using sticky paper to trap these mites makes identification easier for the Varroa mite.

Question - How effective is menthol against the detection and control of varroa mites?

Fred Fulton
Montgomery, AL

Answer - Menthol is used primarily and exclusively for the control of tracheal mites. It does not work well or at all on Varroa mites because of their development cycle and where it takes place. Varroa adults are just too big and robust to be affected significantly by menthol vapors.

Menthol's use and effectiveness can be highly variable. Because the material must "evaporate", its concentration in the air is dependent on temperature, humidity and air flow. If all of these conditions are at optimal levels, tracheal mite control can be as high as 90-95%. If the conditions are not optimal, 0% control is also possible. Every range of control between these two levels is also possible.

STICKY PAPER FOR VARROA MITE I.D.

CHAPTER 15

PESTS AND PREDATORS

WAX MOTHS

I keep receiving questions about what to do about wax moths since there are no "chemical" controls currently approved in the U.S. Here is a reprint of a Jan. 1994 Classroom answer addressing possible wax moth controls. (Jerry)

Dr. Keith Delaplane at the University of Georgia has compiled a list of these alternatives.

1) Leaving combs from honey supers exposed to light and air. The light-colored combs that have been used only for honey may be stored in supers stacked in alternating directions, to allow in light and circulating air. This will not work for darker comb.

2) Freezing Combs. Wax moths, including the larvae or "worms" are easily killed by freezing temperatures. If only a few frames have moths, and if you have room in your freezer, put them in overnight. Then the frames can go back into the hives where the bees will repair the damage if it is not too severe.

3) Use dry ice, which is frozen carbon dioxide. The moths will suffocate if carbon dioxide from dry ice pushes the air out of any enclosed place where they are living. A system that might work well is to stack the supers on a piece of plywood or bottom board that seals the bottom well. Then tape up the cracks between the supers. Finally, put about 10 pounds of dry ice, wrapped in paper, on the top bars of the frames in the upper super and cover it with an empty super and plywood on top. If the whole system is sealed well, the carbon dioxide will build up inside and suffocate the moths.

Be careful when using dry ice. It is extremely cold and will freeze your skin if you touch it with your bare hands. Also, the carbon dioxide will suffocate you as well as the moths if it builds up in the room where you are working. Work in a well ventilated place when using dry ice and handle it with gloves.

4) A bug light with an electric trap may be set up near the comb storage room to catch and kill moths. This can be effective if it's set up in an enclosed area (like a garage or shed) where there are a limited number of moths. If it's set up outside, it may just draw in moths from the surrounding area!

5) Eliminate cocoon nesting places, like the space between the top bar and the wood strip under it where the foundation is held in place. You can attach the foundation to the frames with staples, instead of using the wood strip. Also, solid rather than divided bottom bars may be used.

6) Keep extra frames on active hives. A strong hive will keep moths away, and will clean up and repair minor moth damages. This is a helpful technique if only a small number of frames need to be protected and enough hives are available.

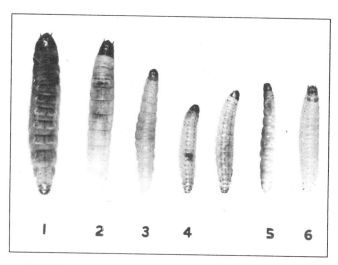

WAX MOTH LARVA AT VARIOUS STAGES

DAMAGE DONE BY WAX MOTH LARVA

HELPFUL HINT:

Dr. Adrian Wenner recently shared his method of getting rid of ants which can be adapted to beekeeping as long as the Diazinon is used well away from beehives.

ARGENTINE ANT CONTROL

Argentine ants have certain annoying habits, among them the following:

1) They may invade houses, particularly when natural food sources dry up and after rains (or watering). Keeping kitchen counters clean at all times helps, but they sometimes go after perfumed substances, including potted plant blossoms. Garbage containers commonly become infested.

2) Ants tend various pest species, including aphids and scale insects. They actually transport individuals up plants, and infest even those within homes. These other insects sap the strength of plants and/or transmit plant diseases. Often their exudate (honeydew) lands on leaves below them and fosters the growth of a black mold, shutting out light from the plants.

An Argentine ant colony in some ways represents a giant amoeba, with its trails similar to the amoeboid pseudopods. By blocking the trails where they enter houses or where they go up trees, one can divert the colony's attention elsewhere. One can follow EPA guidelines and distribute a great amount of poison around the house and garden, thereby polluting the environment to some degree. A far more simple treatment, one not approved by the EPA, will suffice and result in virtually no environmental contamination. A description of that technique follows.

1) Purchase a small bottle of **Diazinon PLUS Insect Spray** (ORTHO, Chevron Chemical Co.) from drug or grocery store. Buy also a medicine dropper, preferably one that comes with a dropper bottle.

2) Save up friction top caps from gallon milk bottles or Kodak film canisters (photo processing centers usually provide them free). These serve as dishes.

3) Find where the ant trail enters the house or starts up a plant. Using the lid or cap as a dish, place into it 5-6 drops of the undiluted liquid poison.

Within an hour dead ants should be seen on the surface near the dish, and within a couple of hours the ant trail will no longer exist. Save and wash "dishes' in detergent water.

Other techniques and information about Argentine ants can be found in the following book: *TINY GAME HUNTING: Environmentally Healthy Ways to Trap and Kill the Pests in Your House and Garden*. By Hillary Dole Klein & Adrian M. Wenner, 1991 (Bantam, paper, now out of print but available in some bookstores).

NOTE: The very small amount of ant poison used in each "dish" poses no problem for pets or children, especially since it has an obnoxious odor. Furthermore, the poison breaks down rapidly in the air and becomes ineffective even against ants by the next day.

Question - I am a new beekeeper and subscriber to ABJ. I have been reading back issues of ABJ and have noticed several articles that indicate that wax moths are killed by frost or freezing. In "The Wax Moth Problem" (Nov. 93 ABJ, Pg. 765 3rd down) Steve Taber writes, "We know that combs and bee equipment generally exposed to light and frost will kill all stages of the moth..." At our meeting of the S.W. Missouri Beekeepers' Association I brought this up and it seems most people here believe this. However, I purchased some used vacant hives this winter. They hadn't had bees in them for about five years, had been sitting up off the ground covered with a tarp, and had been put away with some frames full of combs. They were thick with wax moth cocoons, and many small (1/4" - 1/2" long) larvae had moved under the frame edges. When I brought these hives home, I dumped the frames out of the boxes (Over 500 frames) and scattered them on the ground in my yard. The grass was short, the location wasn't sheltered - they were out in the open where the wind could blow through the frames. They laid in the yard for about a week. The ground froze and on at least one night the temperature reached a low of 10ºF (I have max./min. thermometer nearby). When it warmed up into the 60s, I cleaned up the frames, removed all the wedges and found thousands of live wax moth larvae living under the wedges. Some had burrowed into the wood. I did not notice a significant number of larvae that appeared to have recently frozen to death. Nearly all of the larvae I took time to look at closely appeared very healthy and able to crawl about.

How can this be?

Keith & Rebecca Kropf
Seneca, MO

Answer - You are not the only one to give an anecdotal report of wax moth larva seemingly surviving in below freezing weather.

The literature states the following. Wax moth eggs, larvae, pupae and adults will die when exposed to temperatures of 20ºF for 4-5 hours, and 10ºF for 3 hours and at 5ºF for 2 hours.

I've had letters from northern beekeepers wanting to know how wax moth larvae appeared in the first few days of spring weather. One gentleman from Australia wrote that he had produced a lot of round section comb honey and stored it in his home freezer. When he placeD it in his indoor honey stand, he noticed wax moth larvae crawling around inside his sealed sections. Other stories similar to your come in several times a year.

All that said, I called several universities and USDA bee researches and told them exactly what I was told. The ability of the various stages of wax moth life to tolerate cold has really not been extensively studied. Other insects have been studied to find out the mechanism that allows their survival over harsh northern winters, but not wax moths.

I found out that the eggs of many types of insects can tolerate cold extremely well. This makes sense since most insects use summer to grow to adult breeding stages, mate and then lay their eggs in fall to hatch in spring and start the cycle all over again. Also many insect stages including larva, pupa, and adult have a sub-stance called glycerine in their tissue which inhibits freezing–kind of like antifreeze. If liquid in them does not freeze, then their cells stay intact and life can be sustained for longer periods.

Now the $64,000 dollar question–are any of these mechanisms for surviving below freezing temperatures found in wax moth eggs, larvae, pupae and adults? Or, are the reports of wax moths in various stages surviving apparently extreme cold just mistakes in observations? I don't know.

Anyone have any other thoughts?

Question - I'm finally catching up on reading my ABJ, and I found your comments about controlling ants in the October 1993 issue to be slightly lacking. One of the readers (James Armstrong) commented that he was unable to keep ants out of his hive and that they were building their nest on the inner cover. Your answer to his question about how to get rid of the ants focused on getting rid of the ants and then setting up a grease barrier to the ants so they could not repopulate the hive. The theory is good, but I wonder how well this will work?

I inspected bees in central Ohio for the past two years, and found numerous hives infested with ants. The ants were basically of two types. The first was fairly large, black carpenter ants. The other type was a small black and/or red ant. These small ants infested hives by the thousands, scattered quickly when disturbed, climbed onto the beekeeper with ease and bit!!! I do not think that a hive could be cleansed of these ants without a lot of aggravation and time.

Most of the hives that were infested with ants had inner covers that had no holes, creating a dead space between the inner cover and the telescoping lid. Many of these covers also were not shimmed on the up side, so the telescoping lid fit flatly on the inner cover except when either was warped. In the hives that had holes in the inner cover and that had at least a bee-space between the inner cover and the telescoping lid, ants did not seem to be a problem. Perhaps this is because the bees are able to patrol the area and prevent the ants from setting up a nest.

I would therefore add to your answer to Mr. Armstrong that he ensure that the bees have access to the space between the inner cover and the telescoping lid so that bees can keep the ants out. This plan has worked well in my own hives and in the hives of most of the beekeepers who have followed my advice. It is a nice natural control method that has no worries of environmental pollution.

I have also heard of many beekeepers using various substances on the inner cover to discourage ants from building nests. These include powder cleaners such as Comet, salt, nutmeg, and hot pepper. I do not recommend using these because of the potential honey contamination.

I do not know who sells the inner covers without the holes or the shim, but they seem to be used by many beekeepers in this area. I always thought that the purpose of the inner cover was to provide ventilation at the top of the hive. These inner covers sealed the top of the hive. When I asked people why they used covers of this design, most just shrugged and said that was the way they came. My question to you is, "Is there a good use for this type of inner cover?"

Thank you for providing this excellent column. I have learned quite a bit by reading it.

Christopher Cripps
Columbus, OH

Answer - The ant question and the continuing comments from readers throughout the country has amazed me. This "controversy" has generated more interest than any other since I have been writing the "Classroom" in the American Bee Journal. I appreciate your comment and question also.

About inner covers - I do not know who sells inner covers without the oblong hole in the middle, but I can understand why. I was taught that the inner cover had at least two major uses and reasons for being. The first had to do with ease of hive manipulation. Before the advent of the inner cover, the top of the hive would become glued down with propolis. When telescoping covers came into more use as a better cover from the elements, it was as you may visualize, even more difficult to get a hive tool or other prying instrument up and under the telescoping side to "pop" the top loose. Somebody thought let's use an "inner cover" to prevent this awkwardness in removing the telescoping cover. The inner cover gets glued down, but not the telescoping cover. The second reason I learned was that the bee escape device was invented and its design required a partition between the supers and hive body with the device in the partition. A hole was cut in the inner cover to install the bee escape and voila another use. I have never heard anything about the relationship between the inner cover and ventilation.

So, depending on what kind of cover you use, telescoping or not, and if you use the standard bee escape, this will determine if you use an inner cover at all or one with a hole or not.

Yes, if the colony population is strong enough, they can remove the ants that they can get to.

Question - Controlling the wax moth problem is a big challenge for any beekeeper. So, I am constantly scanning for any news about means to control the wax moth. In an earlier issue you answered someone's question on this subject. In your answer you talked about several methods of controlling the wax moth, each relied on home grown type remedies. I infer that there aren't any powerful anti-wax moth type chemicals on the market. I know of an Eastern European professional beekeeper, who allegedly purchased, in USA, a substance for controlling the wax moth. I am curious as to whether chemicals for wax moth control are available, but cannot be purchased here because of environmental considerations.

I would appreciate very much if you could shed some light on this matter.

Joe Volz
Matawan, NJ

Answer - Yes, there are various chemicals, liquids, fumigants, and bacteria which will utterly destroy wax moth eggs, larvae and adults. Why are they not in use? Because they are either very toxic to the bees, beekeeper or the potential consumer of hive products or, based on the size of our industry in North America (sales), it is not worth it for a company to seek registration (costly).

Beeswax is a natural fatty substance which, because of its composition, is a <u>chemical sponge</u>. Anything which is oil or fat soluble will be grabbed by the beeswax comb and hung on to so to speak. Some chemicals once locked up in beeswax can be liberated from the wax by "airing" out the comb, but 100% of it never completely leaves. There's always some left behind to affect the bees and brood in the colony. If honey is stored in those cells where the chemical is also stored in the beeswax, this chemical may leach into the honey. Then you have a food product that is not pure, adulterated if you will, and probably to some degree a health hazard to whomever is eating it.

I have to applaud the EPA on one hand for making an effort to protect the honey-buying public, but on the other hand chastise them for making it incredibly expensive for safer products to come to the market. Within the last several years there was one wax moth control product that left the market because the EPA asked for additional data for re-registration that would have cost between $500,000 and $1,000,000. It doesn't require a CPA to figure that this was not cost effective for them to do.

At the present time the USDA is working on a natural control method that hopefully will be able to jump all the appropriate EPA control hurdles. Even at this, assuming everything goes perfectly, a product is still several years away.

My only suggestions to you are use the natural control methods mentioned in the January 1994 Classroom for now and don't eat your eastern European friend's honey.

Question - One day after I had been away for a while I checked my hives and found one where all the worker bees had been killed and white-faced hornets had taken over. I opened the hive and to my surprise saw the queen dart to the top of a frame with a white-face hornet in pursuit. A second later they locked in combat and to my amazement the queen was all over her. In about ten seconds the queen had killed the hornet and she fell to the bottom.

Do queens have any special weapons that workers don't have that would make this possible or are they (queens) more vicious?

Larry Werstles
Austin, PA

Answer - As you have seen, queens can and do defend themselves quite well when necessary.

I have a couple quotes for you out of the new edition of The Hive and the Honey Bee which show that queens have the equipment to defend themselves. On *page 378, we find this, "Queens, but not workers, produce functional venom at the time of adult emergency, and the venom sac of the queen contains 3 times more venom than that of the worker." And on page 156, "The sting of the queen is longer than that of the worker and is more solidly attached within the sting chamber."*

Queens generally don't have to use their stingers to defend against hornets, but from the above information, they can put up a good fight.

Question - I had 3 colonies of bees that were overrun by yellow jackets. They took all the honey and drove out all the bees.

They uncapped all of the honey in every single frame. Can I still use these frames? What causes this?

Jim McGrady
New Berlin, WI

Answer - Yellow jackets are opportunistic predators. They feed on other insects and their young just like we feed on cows and chickens for our food. They attack colonies of honey bees that have been weakened by disease, parasitic mites or other reasons and cannot defend the colony. The yellow jackets cleaned your colony out and there is no reason a healthy colony cannot be installed on the equipment left, if the colony did not die from foulbrood.

Question - I am intrigued with the female wax moth's ability to locate isolated colonies of bees, when the nearest known colony may be several miles away. Considering the vastness of even two or three square miles loaded with trees and shrubs and a single lone beehive that a female moth is able to find seems quite a feat. Any idea? Perhaps the wax moth also attacks deserted bumble bee and wasp nests. Any evidence of this?

Another question. What is the lowest thermal limit that wax moth pupae can tolerate before they become frozen? Does the moth occur in northern areas below this lethal temperature? If so, what studies have been made to explain it?

Robert Weast
Johnston, Iowa

Answer - I didn't feel qualified to give you a complete answer so I had to call in the big guns. I forwarded your question on to the foremost authority (I think) on wax moths, Dr. Hayward Spangler at the Carl Hayden Bee Research Center. It's great to have this type of expertise available to the beekeeping industry.

Dr. Spangler: I assume that by wax moth you mean the greater wax moth (GWM) since that pest does much more damage to honey bee equipment in the U.S. than its behaviorally and morphologically distinct close relative, the lesser wax moth. I don't know how far a female GWM will fly in search of food for her offspring. However, these moths are strong flyers and I think they are capable of travelling more than just a few miles. Like many nocturnal insects, they can probably use the moon or a star as a reference to maintain a consistent heading. It is possible that they may fly high enough to be carried long distances by wind, but I tend to think that they fly fairly close to the ground. They have good hearing and should be able to avoid insect-eating echo-locating bats much of the time.

If a hive contains comb but no living bees, the GWM will have difficulty locating it. However, when occupied with bees it is a warm and odorous target that the moths may locate by smell. Perhaps ventilating bees spread more hive odor into the surrounding air, or odors from the bees may attract the moths. The moths can hear high frequency sounds produced by wing-fanning bees some distance from the hive. However, it remains to be demonstrated whether moths locate a bee hive by detecting these sounds.

I know of no reports of GWM infesting wasps nests. However, some of these moths have been taken from bumblebee nests. In general, though, they are only found in the nests of the genus Apis (honey bees).

Freezing temperatures kill all stages of the GWM, eggs, larvae, pupae and adults. For example, within 4.5 hours at 20ºF, all stages die. At colder temperatures the moths die more quickly. Cantwell and Smith (1970, American Bee Journal 110:141) give information about killing greater wax moths with both freezing and high temperatures.

In the U.S. the GWM certainly occurs far north of where freezing temperatures begin to occur. In the more northern areas many moths are no doubt killed by the cold. But in most instances some probably survive in hives warmed by the bee colonies. Clusters of GWM are also capable of raising their own temperature temporarily during larval development. The surviving population gradually builds up large enough to cause destruction perhaps by August. Migration from warmer climates may also add to the late season population in the north.

The GWM is less destructive in the Pacific Northwest and at higher elevations throughout the mountainous west. Low temperatures and terrain may be major reasons for this. In these regions the lesser wax moth frequently causes severe damage. Research is planned to determine if the lesser wax moth is more tolerant of freezing temperatures.

Question - I am a beekeeper in California. I started beekeeping as a hobby. My question is this. I was given two hives by my father. One was very small; the other was a larger hive. The small hive is doing very well. I gave the hive two frames of brood from another hive. The larger and stronger hive is the problem. I went out to check the hive three days after I got them home. I knew there was something wrong as soon as I got to the hive. There wasn't a bee to be seen. When I opened the hive, there was nothing but dead bees. There were about 150 dead bees all around the queen excluder. I also noticed there was brood in the super. The color of the wax was very dark, almost dirty looking. Most of the cells in the center were open. The few I found with larvae (also dead) appeared to be stringy, almost like mucous. The other thing I noticed was a big worm on one of the frames. It was about an inch long sort of a whitish green color.

In a panic I burnt the hive. My concern is that I have 8 other hives that I have raised for 2 years. They are super strong colonies - with very light gold honey, very clean white wax and cappings. I am afraid that whatever these bees had may infest my other hives. Being new to the work of bees, I hope you will forgive my ignorance. I'm not sure what to watch for or what to do. I have never had a problem like this in any of my other hives. I hope you can help.

P.S. - Yes, I learned the hard way, never bring an uninspected hive home.

D. Saul
Hollister, California

Answer - My most memorable lessons in life and in beekeeping I've also learned the hard way. But, you remember the lesson and it can result in improvement for you in the future.

Your description of symptoms sounds like American foulbrood. The bees died, some probably found other hives to live in. The worm is probably a wax moth larvae which feeds on remains left in old comb. You burnt the hive which in the condition you noted was exactly the right action to take. AFB is a very long-lasting and difficult disease to control in its latter stages.

Your concern, rightly, now is how to identify this and other diseases to protect your colonies left. Start a regular spring and fall Terramycin preventative treatment program. I would also suggest that you become comfortable in opening and manipulating your colonies and that you become aware of the healthy condition of bee larvae and pupae in particular. A good reference book is always invaluable. The new edition of The Hive and the Honey Bee at over 1300 pages and only $36.00 plus postage, should be on your shelf. Pictures, descriptions and treatments will be at your fingertips. Also, join and attend your local beekeeping association meetings. You can tap into this reservoir of knowledge and experience while keeping up with the latest in the beekeeping industry and having a good time.

Keep your conscientiousness and your desire to learn and remember Jerry's Law #2. "Stay calm, there will be plenty of time to panic later."

Question - Would you recommend some method of keeping small ants out of the hive? I've tried most every way including putting trays filled with oil under the hive where it rests on blocks. They still get in. So?

Rich Barsdale
Homeland, CA

Answer - First, be sure that these small ants have not bypassed your protective measures and set up their own colony inside your hive, under the bottom board in the lid, etc. Sometimes they will do this.

The most effective ant barrier hive stand that I have seen was a metal hive stand approximately 1 ft. off the ground with each leg set in a can of used motor oil and the rest of the leg smeared with grease. That stopped all ant traffic.

Question - I need some expert advice. I have one hive of bees to which ants are attracted and I cannot figure out how to keep them out of the hive. I find them on the inner cover when I remove the hive cover.
Any suggestions would be helpful.

James Armstrong
Fairlawn, Ohio

Answer - In some parts of the U.S. there are species of ants that are actually predators of honey bees, their eggs, larvae, pollen, honey and use the hive as a nesting site.

These types of predatory ants obviously need to be restricted from honey bee colonies. The method used most often is to be sure all ants, eggs and their queen are out of the hive and then put the hive on a stand having legs lifting the hive above ground level. These legs are set into cans filled with used motor oil. This acts as an excellent ant barrier. Some beekeepers smear the legs of the stand with grease or some other sticky agent that won't wash off. (However, as you have read earlier in this column, some beekeepers are questioning the environmental and food quality safety of using used motor oil and grease to deter ants. If motor oil is not satisfactory for you, try mineral oil and grease the legs with food-approved grease.)

The ants you have in Ohio probably are not the predator type. But they can still be a nuisance. If possible, moving hives to open treeless areas often helps.

If I were an ant, the selection of a bee hive is a pretty good decision. It's relatively warm and dry and protected from any predators plus it may yield some food. Who says they're dumb?

Question - I have lost four hives of bees in the last two years. My trouble is wax moths. Two hives I burned up and two I cleaned up and used again. Was that a mistake. What do I do? I don't have trouble with mites.

W. Breckenkamp
Quincy, IL

Answer - I think you do have either a mite problem (varroa and tracheal mites) and/or a disease problem in your colonies. The reason I say this is that a strong healthy colony is the best defense against wax moths entering a hive. It is only when a colony is weak from other causes that wax moth females can sneak in a hive, lay eggs and leave.

I've included a Dadant catalog for your review. Check the medications and mite control pages.

Don't burn your hives for wax moth or mite infestations. Badly damaged individual frames may have to be replaced along with comb, but there is no reason to burn a perfectly good hive body, cover or bottom board. The only time total hive burning might be advised is with American foulbrood. However, even with this virulent disease, beekeepers can still salvage empty bodies, supers, covers and bottom boards by scorching them with a torch and then repainting.

Question - In Vermont pest control is 50% of raising bees. We have what we call "Earwigs" - insects about 3/4" long with pinchers on their tails held arched over the back. How does one kill them?

Peter Turmelle
Waterbury Center, VT

Answer - Thank you for your interesting question concerning earwigs in bee hives. I had never heard of earwigs in bee hives or of them causing any damage. But, in searching the literature I found that earwigs can be a problem in beehives.

Apparently, earwigs are opportunistic scavengers and will frequent bee hives looking for an easy meal. They can pierce honey cappings and spoil comb sections with fecal material and other food particles. Reports have also been submitted saying earwigs will feed on uncapped larvae.

Earwigs may be trapped by filling flower pots with hay or straw inverted close to the hive. They are attracted to the hay-filled pots. They can and should be occasionally removed and destroyed, especially if they appear in great numbers.

Question - I'm from Canada and have just heard that PDB (Paradichlorobenzene) is no longer available in the States for wax moth control. Some of my friends, especially in your southern states, have a difficult time controlling wax moths. What are their alternatives?

Lorraine Lehman
Spruce Grove, Alberta

Answer - You are right that PDB is no longer registered in the U.S. for wax moth control on bee hives. Our EPA in its infinite wisdom made the re-registration of the product very difficult to justify. As a result, we are now without an easily used wax moth control.

I have heard that some beekeepers have purchased "moth crystals or balls" in retail stores to use. Sometimes this material is a different chemical called naptha-lene which is absorbed by the beeswax and not released. Bees nor queens will use combs with napthalene in them, so now the beekeeper has hundreds of unusable combs. Tell your friends to not make this costly mistake.

There are some alternatives though. Dr. Keith Delaplane at the University of Georgia has compiled a list of these alternatives.

1) Leaving combs from honey supers exposed to light and air. The light-col-ored combs that have been used only for honey may be stored in supers stacked in alternating directions, to allow in light and circulating air. This will not work for darker comb.

2) Freezing combs. Wax moths, including the larvae or "worms" are easily killed by freezing temperatures. If only a few frames have moths, and if you have room in your freezer, put them in overnight. Then the frames can go back into the hives where the bees will repair the damage if it is not too severe.

3) *Use dry ice, which is frozen carbon dioxide. The moths will suffocate if carbon dioxide from dry ice pushes the air out of any enclosed place where they are living. A system that might work well is to stack the supers on a piece of plywood or bottom board that seals the bottom well. Then tape up the cracks between the supers. Finally, put about 10 pounds of dry ice, wrapped in paper, on the top bars of the frames in the upper super and cover it with an empty super and plywood or inner cover on top. If the whole system is sealed well, the carbon dioxide will build up inside and suffocate the moths.*

Be careful when using dry ice. It is extremely cold and will freeze your skin if you touch it with bare hands. Also, the carbon dioxide will suffocate you as well as the moths if it builds up in the room where you are working. Work in a well-ventilated place when using dry ice and handle it with gloves.

4) *A bug light with an electric trap may be set up near the comb storage room to catch and kill moths. This can be effective if it's set up in an enclosed area (like a garage or shed) where there are a limited number of moths. If it's set up outside, it may just draw in moths from the surrounding area.*

5) *Eliminate cocoon nesting places, like the space between the top bar and the wood strip under it where the foundation is held in place. You can attach the foundation to the frame with staples, instead of using the wood strip. Also, solid rather than divided bottom bars may be used.*

6) *Keep extra frames on active hives. A strong hive will keep moths away, and will clean up and repair minor moth damage. This is a helpful technique if only a small number of frames need to be protected and enough hives are available.*

Question - I have enclosed the remains, I think, of some kind of worm that has gotten into my hives. What are they and how can I control them?

Frank Williams
Goode, VA

Answer - Those things you sent are the cocoons of the "wax moth". There are two different species of wax moths, the Greater and the Lesser, but they do the same thing, so we will just call them wax moths.

When a colony is weak from other diseases or parasites, it cannot defend itself from intruders. In your case your colonies were weakened and allowed a female wax moth into the hive. She layed eggs, they hatched and the larvae started tunneling through the comb looking for food. Wax moths cannot live on just beeswax, but they eat the larval skins left behind by developing honey bees, pollen, etc. In the process of searching for food, they destroy the comb. When they are ready to pupate (turn into an adult), the larvae sometimes hollow out a depression in the wood of a frame or hive body, weakening the wood, of course. They then spin a cocoon and an adult emerges in as little as a week in warm weather.

The best defense in a living colony is to be sure that it is healthy and strong. If it is, then these intruders can be kept under control by the bees.

There are no chemical or biological controls currently registered for use in the treatment of wax moth infestation either in living colonies or stored equipment. The use of freezing or carbon dioxide treatment are the only suggested treatments available at this time. Even if empty comb is frozen (either outside or in a commercial freezer) constant vigilance is necessary to prevent reinfestation until those stored supers are returned to the bees.

Wax moths do hundreds of thousands of dollars of damage to honey combs every year in the United States.

Question - My question is in regard to wax moths and honey bees. I understand that the best protection from wax moths is to maintain strong, healthy colonies. however, while looking through another magazine, unrelated to beekeeping, the other day I found a story about bats that included plans for bat roosting boxes and wondered if a couple of bat houses in my apiary could help control the wax moth? It seems to me that since bats and the wax moths both fly at night and my bees don't, that this could be one of those natural relationships that I could take advantage of. Do you have any information on bats and wax moths? Thank you.

Michael Sobotka
Pawnee, IL

Answer - *The wax moth experts at the USDA Lab in Tucson, AZ tell me that bats can and probably do feed on wax moths, but they are not a major predator. The wax moths, when flying, stay close to the ground and other structures to make it difficult for bats to home in on and eat them. The wax moths also can hear the bat's frequency very well on their built-in radar units and fly in zig-zag and loop-the-loop patterns when they hear a bat echo location. An analogy would be, of course, a fighter aircraft trying to evade a missile homing in on it.*

Bats eat a lot of other harmful, destructive and bothersome insects, so if you were planning to put up a bat roosting box, I'd go ahead.

WASPS AS PREDATORS

Question - Last spring I had 12 hives and now I don't have a singe bee. Wasps were so aggressive that they would attack the bees guarding the entrance and wrestle with them until the guard bees were killed or the wasps flew away. Wasps would fly away with the dead bees until all my bees were gone - not even one bee in the hives was left dead or alive. The only thing I did was remove all the queen cells in June as I have done in years past.

Ray Moeller
Crete, IL

Answer - *Sometimes wasps do eat honey bees and will attach colonies and kill several dozen honey bees. Wasps are opportunists and do not want to jeopardize their own existence by attacking a large healthy colony. My guess is that your colonies were* *weakened by queenlessness, mites or disease and the wasps were merely taking advantage of the situation. There could be many reasons your colonies were weakened. I would suggest you call your local bee inspector and have him come out to inspect your hives and hive debris for any disease or parasites that might be identified before you install new bees in your old equipment.*

Question - About 15 of my hives were invaded by what I call "wood bees" this summer. The bees in all 15 hives were destroyed by the "wood bees" and then the wax moths took over and ruined by frames. Do you know anything about these "wood bees" destroying colonies and what can I do to prevent this next season when I start up again?

John Barnes
Indiana

Answer - The samples you sent us were still in good enough shape to be identified as one of the "Vespid Wasps". They are a large species of yellow jacket that are eating your bees. Normally, they feed on a variety of insects or bother people at picnics for a bite of their hamburger or hot dog. They are carnivorous. Honey bee colonies generally are well protected by guard bees and can defend themselves quite well against these predators. Only when the population of these wasps increases unexpectedly at times or a colony is weakened because of mites and/or disease do these predators become a real problem.

I could understand maybe one or two hives being weak and singled out for attack. Having 15 hives attacked is both remarkable and discouraging to you, I'm sure. I'm going to venture a guess that your bees were weak or weakened by mites, disease, pesticide sprays, etc., and were picked out by this healthy population of wasps as a good and easy food source.

The main thing you can do is to be sure that your bees are strong, healthy with good populations. If you see wasps feeding on them, try making the entrances smaller and more easily defended. I wouldn't necessarily go out on a search and destroy mission on these wasps. They do feed on quite a variety of harmful insects. They were just getting an easy meal this time. It just happened to be your bees.

Moving your colonies to a new location would be another option, but this is not always possible or easy to do.

PROTECTING STORED COMBS

Question - I have been keeping bees since 1917 when I was a freshman in high school. They are my great joy.

I have two questions for you. First, what is the best way to keep wax worms from hatching in stored frames? I'm afraid of chemicals. My second question is in 1918 there was a small monthly honey bee newsletter published in Covina, CA called "Western Honey Bee". It discontinued publishing in the war years. I am wondering if some bee man out there would possibly still have a copy of that magazine that I could beg, borrow, or buy. I have a nostalgic yearning to hear of a copy still in existence.

Jack Lowe
Oroville, CA 95966

Answer - Unfortunately, wax worms can only be easily controlled with various chemicals. Pari-dichlorobenzene (Dadant brand "fumigator") is the most widely used and effective one. Several years ago there was a biological control using bacteria to kill the wax worms, but this was discontinued because of EPA regulations and expense running into hundreds of thousands of dollars for new regulations. (Another example of inefficient and shortsighted government help).

Surely someone out there has a copy of "Western Honey Bee" for Mr. Lowe. True beekeepers never throw anything away. Dadant & Sons has the volumes 1919-1930, but they are in our apicultural library so we don't want to part with them.

Question - At one of my bee yards this summer I noticed more than once quite a number of dragonflies in the air. I would like to know if dragonflies kill a significant number of bees, and if so, is there anything to do other than find a different location. I got a fair yield of sumac honey from some of my hives there, so I would not want to move except for some good reason.

Dwight Tew
Franklin, TN

Answer - Although dragonflies are voracious predators of other insects, we have not heard of them being a problem as far as worker bees were concerned. We have heard reports, particularly in the southern areas, where they do take quite a toll of virgin queens on their mating flights. However, we hadn't heard where they had decimated the field population of colonies. One would think that the numbers of dragonflies would have to be extremely high to have any effect. The presence of large numbers of dragonflies in your bee yards may have been due to the presence of the honey and the numbers of insects other than bees that might have been drawn to the honey. The dragonflies may have found your honey-laden hives to be a convenient lure for unsuspecting insects.

WAX MOTH PROBLEMS

Question - My question is of all the research and medications, etc., why hasn't something been developed to take care of wax moths?

Joe Farley
Broken Bow, OK

Answer - *The only way to control wax moths in a live colony of bees is to keep that colony strong by having a young prolific queen. Wax moths in a live colony are a sure sign that the colony is either queenless or in a weakened condition due to disease or some other malady.*

If you are referring in your question to protecting stored combs, this can be accomplished several different ways. The easiest protective method is to keep the stored honeycombs free of brood rearing. Wax moths are attracted to combs that have had brood reared in them before.

Secondly, you can store the combs in outside temperatures in an enclosed mouse proof building. Freezing temperatures will kill the wax moth.

In fact, hobbyist beekeepers who only have a couple of supers store them in the freezer.

If your winter temperatures are warm enough that wax moths continue to be active throughout the late fall and winter, then you need to use some other method of control. Wax moth crystals or paradichlorobenzene as it is called, is the main method now used by U.S. beekeepers. A few still fumigate with carbon dioxide, but this is dangerous and requires special equipment and handling.

The biological moth control called Certan® which was sprayed on the combs is no longer commercially available, so this option is not available any longer.

If you use wax moth crystals, be sure to use the paradichlorobenzene formulation rather than the naphthalene formulation. The crystals need to be replenished at regular intervals or at least checked to make sure that there are still crystals present during super storage.

Also, the fumigated combs need to be aired thoroughly before placing them back on the colonies for the honey flow.

Question - A few months ago there was a question sent in asking what to do about wasps attacking a colony of honey bees. Just to show you how far the arm of the *American Bee Journal* reaches, we received a letter from Stephen Makowiecki of Warsaw, Poland who had read the question. This is his experience and treatment for the problem.

Answer - *I read Mr. Roy Moeller's question regarding damage done to bee farms by wasps with a lot of interest. I would like to add a thought to your answer, however. I have had some experience in fighting wasps and developed an efficient means to defeat them.*

I run a bee farm in northern Poland, a region of a thousand lakes and large forests. Steve Taber, a friend and famous beekeeper, visited me in 1989. It was a very difficult year because of a great number of wasps and hornets! Steve was sorry for me as he watched me battle to protect the bees. I managed to kill about 40 hornets and even more wasps with a swat! I had to run after them like a madman, however, which amused Steve a lot.

Both wasps and hornets come from the same kind of insect, Vespidae, and have similar behavior. Wasps rarely eat full grown bees; hornet do that. Wasps get inside the beehive and rob it of larvae, royal jelly or honey. Larvae are the greatest attraction because they contain clean living protein which is a delicacy to wasps. To them it is what tender beef is to me. I also like beer and I discovered that wasps like it, too. Paradoxically, I found what I was looking for through this shared interest. I placed beer bottles, half cup full, all over my bee yard. The bottles had wide necks; not to scare the wasps, but invite them. Once inside, they couldn't get out! It is best to set out the bottles in early spring when only mothers have lived through the winter. One hornet or wasp killed in the spring is as good as ten of dozens in the summer.

Today I drink beer together with wasps and hornets and with satisfaction, I watch them fall into my traps.

Stephen Makowiecki
Warsaw, Poland

Question - I work for a large bee company in the Tropics. I have read in the bee literature about controlled atmospheric conditions to control wax moth, by increasing the level of carbon dioxide. Has any study been done with the use of dry ice for this purpose. If so, where and was it effective and safe?

Kevin Fick
Kealakekua, HI

Answer - I didn't know myself and research info didn't look like it had changed about CO_2 to control wax moths, so I went ahead and checked with Dr. Keith Delaplane from the University of Georgia and Dr. Mike Hood with Clemson for an update.

They said CO_2 was relatively safe to use. For the beekeeper, obviously you don't want to suffocate yourself in a room with high percentage of CO_2, but it is effective in killing wax moths and larvae.

The critical thing is to have a sealed enclosure where CO_2 can be added to a concentration of 98% for 4 hours at 100 degrees F and a relative humidity of 100%.

This would be hard to do using dry ice as the CO_2 source: One because of the temperature problem and two, it would take longer to reach a 98% concentration in your room using dry ice rather than other sources of CO_2.

Question - I am wondering if you have any solutions for ants getting into hives that will not hurt the bees or the honey.

Charles Maxwell
Kuttawa, KY

Answer - Most species of ants in a colony are more of a nuisance to the beekeeper than as a threat to the bees. A healthy colony of bees will keep most ants away from the colony itself and its food and brood areas. Ants being opportunists will take advantage of honey, pollen and sometimes brood if left unprotected by a weak colony. The colony will be weak not from the ants but generally from other causes. The ants are looking for a nice, dry and warm protected home and a bee hive is a perfect place. Many places in the South have predatory "fire ants" which actually seek out and feed on everything edible in the hive. These ants are more than a bother, they are dangerous to all colonies, weak or strong and also to the beekeeper.

Chemical control of ants can be a problem. Any chemical which will kill the ants can obviously also kill bees. If any poisons are used outside around the hive, ants may track the insecticide inside into comb and hive parts before the material actually kills them. Where ants are a problem, hive stands may be used and the legs coated with a product called "Tanglefoot" to keep ants out.

Reader Suggestion - I have read several articles, and have seen questions recently on ants in bee hives and have some observations that might help. I have 5 hives and help several guys with their hives.

Here in the Southeast, where the fire ants are always a bountiful harvest, we have to watch constantly for these pests. When I move or make a new hive, I use 2 cement blocks and cover them with a sheet of roofing felt (tarpaper) and use salt around the area. With an existing hive I lean the hive to one side, just enough to slide the paper under one side at a time.

An entrance feeder is a sure way to attract ants to your hive. To combat this I have taken a piece of 3/8 inch plywood and cut it to replace the inner cover. I cut 2 holes in it so a feeder lid will fit in it snugly. This gets the syrup above the cluster and away from the entrance. I use an empty super to protect the jars, with an inner cover and lid on top.

These steps, along with watching closely for problems, have worked for me and are environmentally safer than other methods I have read about. Thank you for a great monthly magazine.

Joseph Bodnar
Petal, MS

Question - A friend tells me that he combats ants by carrying a plastic container of diesel. He pours a small amount, possibly a tablespoon, down the crack between two colonies on a pallet. The diesel seeps downward, kills ants, and incidentally protects the wood from dry rot. What do you think?

Alan Buckley
Modesto, CA

Answer - *Concerning diesel fuel and ants, I personally am a little gun-shy about endorsing it. I'm positive it would work, but I was taken to the shipping post a few years ago when I told a reader that to protect his hives against ants, he should build some four-legged metal elevated stands, and put each leg in a small can filled with old motor oil or diesel fuel.*

This reader accused me of advocating environmentally destructive tactics by exposing the ground and air to these chemical products. To cut to the chase, this basic contention that there may be a better way to control ants without using these products is probably right. The use of borax, salt, Tanglefoot, vaseline, etc., might be better choices for controlling ants.

CONTROL OF WAX MOTHS

Control of wax moths in empty supers is another problem we have. The chemical Paradichlorobenzene has been reapproved for use in empty supers. Being a carcinogen, its length of stay on the market in a nonregulated form will probably not be long. Some beekeepers after extracting the honey from the supers put the supers back on their colony over a queen excluder and let the bees themselves guard the combs. Beekeepers have also tried freezing and use of CO_2 in a sealed room. Both work, but because of the need for the added equipment (sealed rooms and CO_2 or large walk-in freezers) they are out of the range of possibility for most hobby bee-keepers. The short answer is we do not have a safe, reliable and cost effective method of keeping wax moths out of empty supers. If you do, please share this information with our readers.

Beekeepers living in climates with long, cold winters are having good luck with simply storing empty honey supers outside where the freezing temperatures kill the moths and their eggs. These supers, however, must be protected from moisture and mouse damage. Also, they should be checked weekly for signs of wax moth damage once warmer temperatures return. Perhaps the easiest answer is simply to return them to the hives for wax moth protection once these warmer temperatures return.

CHAPTER 16

PESTICIDES

Question - I have a swarm in the side of my house. It's been there for two years. I have killed a pile of them with Sevin. The pile is about 4" high.

My question is, would it be sensible or profitable for me to now transfer them into new equipment for myself, and how do I do it?

Answer - I think you should forget trying to salvage the colony that you've already decimated with the insecticide Sevin. Having sprayed them with Sevin and already accumulated a pile 4" deep of dead bees, the chances of the queen not being affected or still being alive are not good. If the queen is dead, then you would have to purchase a new one, and if she has stopped laying because of pesticide exposure, she'll also have to be replaced. Plus, what effect has the Sevin had on the remaining workers? I'm not confident of salvaging this colony. You might as well finish killing off the colony, remove the insecticide contaminated comb with stored honey and pollen and burn the whole works. Burning will stop any other bees from being exposed to the Sevin residue. After going to all the trouble of removing the bees and their comb, be sure to plug or patch any remaining holes or cracks in the house that would allow new swarms to enter. The smell of the honey and beeswax could easily attract a new swarm unless you take measures to prevent this.

HONEY BEES KILLED FROM PESTICIDE EXPOSURE

Question - I want to ask you about a problem with combs that have been treated with Vapona strips in storage for over the winter. I put these combs on colonies the next summer and they all died. Now, what can I do with the wax?

Augustin Jordan
Puerto Montt
Chile

Answer - Vapona is highly toxic to virtually all flying insects. It is commonly used in this country and others in a plastic strip (No Pest Strips) that is hung in areas with flying insect problems. The active ingredient in Vapona which as you know is Di-Chlorvos vaporizes into the surrounding air. As the flying insects "breathe" this product as they fly through these pockets of concentrated insecticide, it kills them.

At first glance, this sounds like an excellent treatment for wax moths in stored combs. The problem is that beeswax is a chemical sponge. Beeswax will absorb many chemicals and hang on to them, reluctantly releasing them. This is what happened to you. The Di-Chlorvos was absorbed into the beeswax comb and released at a rate that killed the honey bees that were exposed to it.

To remove the majority of the Vapona from the wax, the wax must be melted in open vats and thoroughly stirred or mixed so that a large surface area of the melted wax is exposed to the air. This will allow the Di-Chlorvos to be driven off the melted wax into the atmosphere. The melted wax should be circulated in this fashion for a minimum of two hours.

Question - In our county there is a mosquito abatement program which sprays Malathion up and down the roads between one hour after dark and dawn. It is dispersed as a fog.

I have been told that this doesn't hurt honey bees as they are not flying. Our hives are located about 50-60 yards from the road.

I haven't noticed a lot of dead bees, but the bees aren't doing as well as I had hoped for either.

What is your opinion?

Edward L. Jones
Monroe, Utah

Answer - The new edition of "The Hive and the Honey Bee" has a good chapter entitled "Injury to Honey Bees by Poisoning".

Suffice to say, Malathion is highly toxic to honey bees when applied to them. Whether 50-60 yards is enough I don't know. Wind, humidity, temperature and other factors would have to figure into the whole scheme.

Personally, I would move my bees if given the opportunity.

Question - While spraying an insecticide for tent caterpillars, many of my supers and brood frames were inadvertently contaminated. The product is a microbial insecticide containing *Bacillus thuringiensis* and the label claims that it is non-toxic to humans, pets, and beneficial insects. However, I am concerned that when these frames are used that the bee larvae may be affected by the insecticide in the same manner as the target insects. It would be cost prohibitive to replace the wax and I was told that a similar product was once sold to control wax worms. Will this insecticide harm my bees in any way? Thank You.

Albert Morgan
Austin, TX

Answer - Bacillis thuringiensis (BT) will not harm any stage of honey bee. A product called Certan® was sold many years ago for the control of wax moth larvae. This product contained BT. It was a liquid that was sprayed on empty combs. When the wax moth larvae started eating, tunneling through the comb, it would ingest the BT which would start multiplying in the larva's gut. This "germ" would destroy the wax moth larva's digestive system and thus kill it. BT is rather selective to moth and butterfly (Lepidoptera) larvae. It will not hurt honey bee larvae or mammals.

I think your frames should be safe for use.

INSECTICIDES

Question - I have three colonies of Midnites in my backyard and I have had the bees there for twenty years. In one of my hives the bees have been dying by the thousands (I think that number is no exaggeration). Earlier they were making a good start with a good queen but the queen disappeared, presumably killed by the same thing which is killing the workers.

I have never had this experience before and I wonder if it might be Nosema. If it could be, can anything be done about it? The other two colonies are normal.

Answer - From your letter it sounds as though your bees are being killed by insecticides, but only a laboratory test can tell you for sure. The Agricultural Research Service, Bee Research Division, United States Department of Agriculture, Beltsville, Maryland, is equipped to make such tests. About 50 of the bees which are dying or dead, but not dried up, should be shipped to them for test. They can also test for Nosema, but I doubt that this is the cause of your trouble.

Question - How can you keep bees when every acre around you is sprayed by plane three or four times a year?

Answer - You certainly have an excellent question concerning keeping bees in an area in which the ground is sprayed by plane three or four times a year. In Illinois, the farmers are supposed to let the beekeepers know when any such spraying will take place in order that the beekeeper may either move the bees out until the spraying is completed or take protective measures. Sooner or later this question of indiscriminate spraying will have to be resolved in order to protect the bees from being completely wiped out. Agriculture cannot allow this to happen.

As I mentioned above, one method is to pick up the colony or colonies of bees and move them from the location to a different area until the spraying is finished. This is dependent upon the farmers notifying you in sufficient time to do this. Another method is to cover every hive with dampened burlap sacks. This should be done the evening before or very early the morning of the day that the spraying is to take place. These sacks should be kept dampened if at all possible and should cover the entrance so that the bees cannot get out. Another helpful practice is to put on top of the top super an empty super shell and then put the lid on top of this. This gives the bees room to cluster while they are being confined to the hive.

NECTAR FROM SPRAYED PLANTS

Question - If the bees gather nectar from sprayed plants, what would be the effect on honey stored in the combs?

Answer - *It is questionable whether much of the poisoned nectar would ever reach the surplus honeycombs. Insecticides applied to non-blooming plants are seldom harmful except when the spray or dust drifts into the hive entrance or onto blooming plants in nearby fields. In the case of early fruit bloom, such as apples some bees might be killed directly in the field leaving very few to make it back to the hive with a poisoned load of nectar. As bees must take the nectar into their bodies, they would usually die of any poisoned nectar before they could return to the hive. Since there would probably be a large portion of the field population lost, there would be little if any surplus honey stored and what honey was stored would probably be placed in the brood area to rear replacement brood for the lost field force. In the case of apple bloom, if the orchardist waits to apply the caly-cup or petal-drop spray as recommended when 75-80 per cent of the blossoms have dropped, there will be little killing of honey bees.*

INSECTICIDES

Question - Our tobacco-farming neighbors (Zebulon, N. Carolina) use insecticides. Will they harm our bees?

Answer - *We have no way of knowing about the particular insecticides your tobacco-farming neighbors will be using but, to be honest, yes, there will be some harm and some losses. If spraying is done properly, there should be a minimum amount of loss to the bees and the beekeeper. Sometimes a part of the field force of the colony is lost but this is usually replaced in a short while by the emerging bees within the colony. Try to get your neighbors to cooperate with you by telling you the kinds of insecticides they will be using so you can become familiar with the effects and try to get them to warn you in advance of the times they are going to be spraying. You will then be able to confine your bees for a short time until the major effects of the spraying have evaporated. You might find that a gift of honey now and then to your spraying neighbors will increase their cooperation with you.*

GARDEN INSECTICIDES

Question - Is there an insecticide for use on gardens that will not harm bees? Our major problem has been aphids on beans and white worms in the ground that eat radishes, etc.

Answer - Most of the commercial seed and nursery companies issue catalogs which carry a number of different kinds of spray material to use on gardens. Some of these are listed as suitable for use on garden vegetables for human consumption. Rotenone, for example, is suggested for control of a wide variety of chewing and sucking insects. Rotenone is considered to be relatively nontoxic, and can be used around bees with a minimum of injury.

DEAD BEES IN FRONT OF THE HIVE

Question - Two of my colonies show good strength and good vitality, but there is an enormous amount of dead bees in front of each beehive? Could you tell me the reason?

Answer - We would have to have more information before we could give you the reason for dead bees in front of a hive. We would suspect spray poisoning such as an insecticide used in areas where the bees were foraging. Check the area for any recent spraying and also send a sample of about 50 bees which are outside the hive but not yet dead to the Agricultural Research Service, Entomology Research Branch, Beltsville, Maryland, for tests.

EFFECTS OF INSECTICIDES

Question - I have a hive of bees nearly wiped out by insecticides. There are nearly 20 colonies in the yard but to date only one has been noticeably affected. Why? What was in the field spray that only one hive was affected? A neighboring field was sprayed by a plane but I would have thought that all of the hives or none of the hives would be affected.

Answer - When we got into the breeding of bees for pollination purposes, we immediately noticed one very apparent fact. Honey bees might be located in the same beeyard but they might forage from entirely different plants in different locations. In California where we were able to follow this foraging behavior, we found that some of the colonies would be foraging, for example, due north into a alfalfa field, while other colonies were flying due west into a safflower field, etc. Where spraying with insecticides is used, it can easily happen that one or two colonies in an apiary would happen to be flying into the insecticide area and thus be hurt by the insecticide while other colonies were foraging in safe areas and escaped damage.

PREVENTING LOSSES FROM POISON SPRAYS

Question - I believe I have devised a plan to prevent bees from suffering a major loss from poison spray. I have one yard of bees about 1 1/4 miles from a commercial peach orchard where they spray, causing a heavy loss among my bees in past years. This year

I fed the bees heavily with thin syrup and gave a cake of pollen before the spraying took place and they suffered very little. I figured the bees mixed the new nectar with the syrup fed and made the poison less potent and at least protected the young bees and brood. Of course the weather was also cool which helped to cut down on bee flights.

Answer - Your experience pretty well parallels our own. Three or four years ago we had bees down in central Florida that were badly hit by a spray used on sweet corn. We found, as you did, that if we kept pollen on the colonies and a continuous feed of thin syrup, that the effects could be pretty well nullified.

INSECTICIDES

Question - I am a small commercial beekeeper and have had no experience with insecticides. At the present time I have 100 hives on S-1 clover and vetch in the Red River bottom in Arkansas. They are within 3/4 mile of where cotton is planted and will be poisoned. What are the chances of getting my bees killed, and to what extent? Are there any precautions to take to minimize the loss?

Answer - It is quite possible that your colonies of bees could encounter the poison as it is applied to the cotton fields in your vicinity. A lot would depend on the relative attractiveness of the cotton blossom as they are being poisoned, as opposed to the clover and vetch which the bees might also be working. My guess would be that they would prefer the clover and vetch to the cotton, so damage to field bees might be minimized. However, the possibility of insecticide drift on to other fields where the bees may be working, and across the yards themselves, must always be taken into consideration. You would know more about the prevailing winds in relation to the cotton fields and your hives. Also, it would make a great difference whether the fields are sprayed from the air. If you can convince the cotton growers to use ground application, and preferably late in the afternoon or early evening hours, your chances of being severely damaged would be minimized.

Question - A friend of mine has a colony of bees that hasn't capped any of the brood. Some of it is in the advanced pupal stage, some nearly full grown and still alive. Some of the oldest are dead. Some are bluish and some a dirty brown. It doesn't seem to be at all ropy, and I couldn't smell anything bad. I cut out a piece of it about 2" x 3" for the inspector to send in to the State College.

After looking all through my bee books, I called him up, and asked if anyone had done any spraying up there. He said the Power Co. sprayed their right-of-way with brush killer. I suggested that he kill the queen and give them another one. What do you recommend?

Answer - Our suggestion would be to send a sample of the bees to the Bee Research Laboratory at Beltsville, Maryland. They should be able to tell you whether they have been killed with an insecticide or a disease.

CHAPTER 17

POLLEN AND POLLINATION

**HONEY BEE WITH A FULL
POLLEN LOAD COMING IN
FOR A LANDING**

POLLEN DRYING

Question - At what temperature should pollen be dried and what is the best method to do so?

Luis Guillermo Clavijo
Bogota, Colombia

Answer - Bear in mind that a pollen grain is the male reproductive element of a flower. As such it carries a genetic code that, if fertilization takes place, may result in a seed with part of that genetic information ready to be expressed. This genetic code is carried by chromosomes and genes which are specially grouped proteins. As you probably know, all proteins whether in meat, eggs, beans etc., can be changed by heating, i.e. cooking. Therefore, you want to dry pollen without heating it excessively and avoid degrading the pollen from its normal most desirable state. My suggestion is to combine low heat, between 100º F and 120º F with a dehumidifier in a closed room. In this way moisture can be driven into the surrounding air and be collected by the dehumidifier without changing or altering the pollen significantly. This method will preserve the proteins, vitamins and enzymes in their natural, but now dehydrated, state.

DRONE ESCAPES FOR POLLEN TRAPS

Question - My bees bring in more pollen than they eat and I have ordered a pollen trap. I plan to install it on the first of June when I'll combine my two-queen colony into one at the beginning of the major nectar flow (wild blackberries) when I'll also install extracting supers. My question is what about the drones? The bees can exit through the trap all right, but the drones can't, can they? I can install escape cones for the drones, but won't they clog up the trap trying to get back in?

Since this is a single backyard hive, I could open the trap each afternoon to let the drones back in. How do commercial beekeepers who collect and sell pollen cope with this problem?

Thank you for your excellent column.

Dan Hendricks
Mercer Island, WA

Answer - Drones are always a problem when dealing with pollen traps or queen excluders. Many pollen traps have special drone escapes which work reasonably well to lessen this problem. If the pollen trap that you have purchased does not have this feature, it may be a good idea to add several drone escapes. Since this is a back-yard hive, you could also do as you suggest and open the traps in the afternoon in order to let the drones back in.

You may want to alternate days that the trap is operational anyway for the following reason. Honey bees bring in pollen in response to the presence of larva that need feeding. So, it is a good sign when pollen is coming in. Honey bees bring in more pollen than they can use immediately as an insurance policy if pollen is suddenly no longer available. Pollen is the honey bee's only protein-source. If you restrict the quantity of pollen coming into the hive for more than a day or two, then other bees in the colony will be "recruited" to stop collecting nectar etc., and will begin collecting pollen. Keep this in mind.

Question - Maple trees in our woodlands are such an early and prolific source of pollen that my bees pack whole frames with it. This surplus focuses on early spring in the phase of light nectar flow. Most of this olive green pollen is not used, nor do the bees ever remove it to make way for honey storage. Therefore, I have had to destroy pollen-laden frames almost annually and replace them with frames of open cells. As the gypsy moth advances in our eastern woodlands, maple is likely to increase its proportion in the forests as oaks reduce, only to frustrate further this problem of pollen abundance.

What are the methods and techniques of thwarting the bees from packing frames with too much maple pollen, particularly in early spring before heavy nectar flows?

Will Johnston
Winchester, VA

Answer - *Maples are one of the earliest and most reliable sources of early pollen in most parts of the U.S.*

Since pollen is the protein/vitamin/mineral source that is necessary for the surge in brood rearing at this time of the year, your bees being the opportunists they are, are gathering as much as they can while they can. The same thing happens with honey. They take advantage of an abundance of nectar and actually store too much for their own needs, which is okay with us because it has value for our needs, whereas pollen stuffed in cells doesn't.

You can't change the bee's genetically based programming to collect as much pollen as possible. But, you can use this ability to your advantage. You may want to start collecting the pollen from the bees before they get into the hives by use of any of the pollen traps offered on the market. You could then control the amount of pollen deposited in the hive. There is quite a market for cleaned, dried pollen as a health food item. Or you could save it yourself to feed back at times of the year when perhaps pollen is not in abundance. Give it a try.

Question - What do we know about bees legs in pollen traps? I started using a couple of front-mounted traps this year and found a few legs in with the pollen. I speculated that they came from dead bees being pulled out of the hive by "undertaker" bees which came off and fell through into the pollen drawer.

Then, the last week of August and early September I suddenly encountered lots of legs. With many summer foragers dying off, they still could be from dead bees. Would a bee actually pull so hard squeezing through the five square per inch (actually 36 in seven inches) mesh that she would pull her own leg off?

If a live bee actually lost a leg in a trap, it would be devastating for her, wouldn't it? Cause her to die, maybe?

Dan Hendricks
Mercer Island, WA

Answer - *The legs that you are finding with pollen pellets are probably parts and pieces from dead bees in the hive. I don't think a live bee would normally pull its own leg off. If she would, with the quantity of legs you saw, you would see bees struggling and caught in the mesh of the pollen trap. If a worker did pull her own leg off, she would soon die as honey bees do not have a blood/body fluids clotting mechanism as mammals do.*

Question - How do I make a pollen patty? Keep up the good work.

Colby McMullen
Williamsville, VA

Answer - *Thank you for your letter of support. I appreciate it.*

How do you make a pollen patty? Pollen, as you know, is the best source of protein and most vitamins and minerals for honey bees. Collecting it and preserving it in small quantities for use later is relatively easy. For use in feeding a lot of colonies (100+) the time, effort, equipment and expense to take pollen from the flower to a patty could be very expensive. The pollen would have more value to sell as a nutritional substitute for humans than its value to feed a colony of honey bees. That's why most recipes for pollen supplement diets do not use pure pollen only, but mix in either soybean flour, Wheast and or Brewer's yeast to the mixtures.

Here are two recipes for a pollen supplement mixture that nutritionally will boost brood production in your bees.

A		B	
Soybean Flour	3 parts	Wheast	3 parts
Pollen	1 part	Pollen	1 part
2 to 1 sugar water		6 to 1 sugar water	

Mix either recipe until a dough-like consistency is reached and form into 1/4 pound patties.

POLLEN SUBSTITUTE

Question - I am interested in feeding a pollen substitute to my bees and have not found a suitable recipe. I've heard that soybean flour can be used, but I don't know where to get it. We had a very cold, wet summer and the bees were not able to store enough honey or pollen. I believe some pollen substitute will be needed for spring buildup.

Clarence Fath
Elkton, S. Dak.

Answer - *To find a "food" that would stimulate egg laying by the queen and sustain the development of larva has been a continuing quest. You would think that by now a perfect man-made bee food would have been found and offered to the beekeeping community, but it hasn't. Bee-collected pollen is the only food that has been found that will both stimulate egg laying and allow larvae to develop normally over a long period of time. That doesn't mean that a man-made food can't stimulate egg laying or allow larva to develop, but none has been found that will do both these things over a few weeks of time after that egg laying decreases and larva begin to die. So, the key is to be able to judge when to put on a pollen substitute in order to encourage early egg laying and brood development just a few weeks before the first pollen is available. If you can do this, you will have the best of both worlds. Remember, honey bees can do no serious colony expansion without pollen. Try this recipe for a pollen substitute patty. (Soybean flour in 100 lb. bags is available from Dadants for about 50 cents per pounds.*

Soybean Flour	3 parts
Sugar	3 parts
Water	2 1/2 parts

(Add sufficient water to form a doughlike consistency)

POLLEN TRAP TRAY READY FOR COLLECTION

Question - I'm a beginning beekeeper (as a hobby) in my second year. I have six hives in the Temecula area of southern California. I wrote to you regarding Apistan strips and you gave me very good advice.

It seems that the beehives are all in good shape and healthy. My beekeeper friend gave me pollen traps and I want to use them, but before I do I need some good information or suggestions as to when, and for how long to keep them on. I don't want to put my bees in jeopardy.

I am aware that I will reduce my honey crop (or will I?). I would rather have the pollen than the honey.

Drago Nizetick
Lakewood, CA

Answer - One can certainly enhance the amount of money made from a hive by collecting and selling pollen–if a solid market exists nearby. However, you may be planning to consume it all yourself or give it away.

As a review of honey-bee nutrition, remember that there are only two foods collected by honey bees—pollen and nectar. Nectar is the sugary liquid produced by some flowers to lure honey bees in for pollen transfer. Nectar is the energy (carbohydrates) source of the colony or when it is converted to honey. Pollen is the source of protein, fats, vitamins and minerals for the colony. As an analogy to the human diet, pollen is the meat, vegetables and fruit; nectar is the dessert. As you know, we cannot live very long on dessert before losing strength and getting sick. We need the meat, vegetables and fruit to help make muscle, bone, nerves, etc. or we die. Even though honey bees are insects, the same conditions must be met in their diet. They need some pollen to provide the nutrients to raise more bees, to collect more nectar to live through the dearth of winter. Without pollen, you have no colony growth.

When you collect pollen, it goes into your pollen-collecting tray not into the hive. The bees realize this and more bees are recruited to stop collecting nectar and start collecting pollen in hopes of bringing in enough to satisfy colony needs. Without some help from the beekeeper, colony health and population will decline.

Depending on the area you are in, you can collect a lot or little pollen. It all depends on your location to pollen-producing plants that honey bees can access. You will have to experiment, but a good rule of thumb is to operate your pollen traps on alternate days. Put the screen or plate used to scrape the pollen off the bees legs in place one day and switch it back the next. This requires more attention on your part, but it assures you a health colony in the long run.

COMMERCIAL POLLINATION

Question - I was told to write to you about how commercial pollinators in your country move their bees from location to location.

I very much appreciate your interest.

Jose Maria Giralt
Argentina

Answer - In this country honey bees are moved from pollination contract to pollination contract on large semi-trailer trucks holding several hundred colonies per load. Generally, the colonies are placed four per pallet and securely fastened with steel or nylon strapping to the pallet. These pallets are then loaded on and off the trailer with specially designed forklifts trucks that can maneuver on soft ground. While being transported, the hives are covered with a net covering the whole trailer load and trucked at night, if possible, and unloaded at night.

This is very hard, unglamorous work, but can be financially lucrative for the efficient operator. Best of luck to you.

Question - I have about 25 acres of apples and I don't seem to get as much pollination as I would like.

I have a commercial bee man who keeps one yard here. He over-winters about 16 hives and splits them about a week before the apple bloom. The problem is the bees start flying only a week before apple blossom and don't have population built up yet.

If I have my own colonies, can I start feeding them a month before bloom and get more bees early?

Mike Robinson
Olympia, WA

Answer - You are basically correct in your judgment that after splits are made there do not seem to be as many foragers. The workers who may normally be foragers are needed back at the hive (split) in order to regulate temperature and feed and nurture the newly developing brood. For effective and efficient pollination to take place, you need good weather temperatures and healthy hives with a queen laying well and lots of foragers ready to go out to collect pollen for developing brood. If you don't have this, then pollination will be less than it could be.

I think that it is only prudent from a business position to specify a minimum of six frames of open brood with bees covering these frames. If you can't get that, then he should bring in double the amount of hives at Apple bloom time. Anything less than that and it isn't a fair business transaction.

If you have your own colonies, you can prepare them so as to meet most of the criteria we already spoke of.

POLLINATION OF RED CLOVER

Question - I am going to produce red clover seed and have been told that I need to get bees with "long tongues". I would also like to know how much increase I can expect from pollination with honey bees.

Answer - A long-tongued bee is not at all a necessity for red clover pollination. It is true that the nectar of red clover is down a considerable distance in the corolla tube; however, the pollen of red clover is immediately available at the mouth of the corolla and honey bees will work red clover quite well. In general, for effective red clover seed production, you would need about 4 colonies of bees per acre. These bees should be scattered in small groups of 6 to 8 colonies spread throughout the acreage rather than concentrated at one side of the field. If other factors are equal, then the use of the honey bees for red clover pollination should increase the seed yield 8 to 10 times or more above normal.

POLLINATION WITH ALFALFA LEAFCUTTING BEE

Question - We read that alfalfa seed growers have increased their number of the pollinating alfalfa leafcutting bee, (*Megachile rotundata*) thereby reducing the required need of honey bees. Now just what is this leafcutting bee?

Answer - The Megachile rotundata leafcutting bee is a small bee that is "a cousin" of a honey bee. It uses the alfalfa leaves to chew up and pack to fashion a nest. It then provisions this nest, which would contain one egg, with a little ball of pollen, normally from the alfalfa flower. Because of this, it is a very efficient alfalfa pollinator. The big problem with this, and other wild pollinating bees, is that it is very difficult to control the populations. Man has learned how to propagate them somewhat, but we still aren't able to cope with the natural disasters that take their toll of them including disease. Because of this, the alfalfa seed industry is convinced that the honey bee is the route for the future.

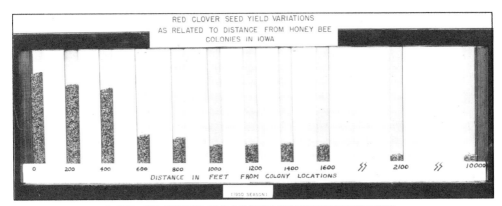

RED CLOVER POLLINATION BENEFITS

POLLINATION

Question - I am going to plant a vegetable and fruit garden in an enclosed area, which will have small window screens all around it. Can a vegetable garden grow semi-protected like this, or do the bees have to get to it?

Answer - The planting of some vegetables that you might anticipate in your screened garden will be satisfactory without bee pollination. This would be especially true of the root vegetables. The only reason you would need bee visitation for them would be if you were planning on saving seed from them.

There are other vegetables such as cucumbers and melons which would require bee visitation to carry the pollen from one blossom (male) to other female blooms. Without access by the bees, no fruit would be set. Depending on the thickness of the screen, some of the smaller insects and beetles may be able to do at least a partial job, but bees have assumed the major role in pollination over much of the country.

There are other vegetables that can produce at least a minimum crop without bee visitation but are aided or helped by bee visitation.

Question - A neighbor wants to pollinate winter grown cucumbers in his greenhouse. We have put in two hives of bees to do the job. Greenhouse temperatures range from about 60 degrees at night to 90 degrees during the day. The ceiling is continuously damp. Our problem is to get the bees to move about freely to work. The bees in one hive refuse to leave the hive; those in the other hive come out but hit the ceiling before they can get oriented. So, all we have are wet wings and bees slithering down to the side of the building.

Answer - *We have had good results with bees in a greenhouse when they had access to the outside as well as inside. We placed the colony outside of the greenhouse with a small opening into the house and the main entrance outside. In this way the main flight of the bees was outside with only a limited number going into the greenhouse. We used this system on tomatoes several years ago. We did use strong colonies since they seemed to give us the best results.*

Question - A few farmers (small scale) in the area mentioned that they noticed bees of many kinds on their corn this summer. My question is this - should I try to rent my hives out for corn pollination? Since I have no experience in this field, I would appreciate any pointers you could give me.

Colby McMullen
Williamsville, VA

Answer - *Corn is one of the large number of members of the grass family. This means that the pollen produced is spread to the female flower parts by air currents. The pollen is dry, large and balloon-like to catch air currents. It floats around hoping to bump into the right place. Honey bees will try to collect corn pollen if there is nothing else available. The pollen being non-sticky is hard to collect and they just knock it off into the air. They do not transfer it down to the silk, themselves, so no pollination can be directly attributed to them. No one rents honey bees for corn pollination. But, if you find a corn grower who doesn't know this and will pay you $50.00 a hive, you can be honest and tell him or take the money. Your decision.*

Question - I happened to see your "Classroom" of the *American Bee Journal*. I am not a beekeeper. I have a 6 acre hobby farm and plant many strawberries and fruit trees. I rarely see many bees which would help pollination on my property.

Would it be beneficial to buy bees and locate a hive on my property solely for this purpose? I'm not interested in extracting the honey since I'm led to believe that the clothing and all equipment would be costly.

Is this practical and can I get the bees from your company?

Richard Mlodzik
St. Joseph, MN

Answer - *As you are well aware, the more insects you have visiting your fruit blooms and transferring pollen from one to another, the better result in more fully pollinated and hopefully better shaped fruit.*

Honey bees have been used for this job because of the great numbers of these insects that you can direct in a given area. Because of pesticide use and the devastating results of mites on feral (wild) honey bee colonies, I am hearing reports from across the U.S. about the lack of honey bees as pollinators now where there were plenty freely visiting flowers just a few years ago.

It would be beneficial to locate honey bees on your property at bloom time for pollination purposes. I am going to try to dissuade you from keeping bees your-self because, as your letter states, you are not really interested in bees or beekeeping, but fruit and fruit production. There is nothing wrong with that. You don't have to be a beekeeper to access honey bees.

What I would suggest is that you contact Mr. Blane White, MN Dept. of Agric., 90 West Plato Blvd., St. Paul, MN 55107, Ph. 612-296-0591. He will be able to put you in contact with beekeepers in your area. What your goal is, is to find a beekeeper who would like a new location for some of his colonies which you would provide. You would receive pollination benefits and he would receive hopefully a good honey-producing location. If having a colony located full time on your property is not possible or wanted, then I suggest you contract for yearly pollination services with a beekeeper.

I'll send you a Dadant catalog and an American Bee Journal. Maybe, just maybe, that spark of beekeeping may touch you!

Question - 1) When bees are used for pollinating cucurbits (cucumbers), apples, etc., do they collect any nectar in the process or are bees rented out for pollination a total loss with regard to honey production?

2) Do Western New York grapes require insects for pollination?

Mark Anderson
Bergen, New York

Answer - Most but not all flowers having both male and female parts on the same flower generally yield some nectar and, some having separate male and female flow-ers, also run the extreme from no nectar to liberal supplies. In other words, there is variability of nectar yield and pollen yield based on the plant, the species of plant, soil fertility, rain, temperature and sunshine.

There are no black and white answers in nature, but there are some generalities which may be used cautiously.

You asked about apples. Bees do gather nectar from apples in early spring. Most times this nectar and apple pollen are being used as fast as they come in because this is the time of year when the honey bee colony is trying to increase its population. So, this early feed source is used to raise brood. If the beekeeper is fortunate to have strong populous hives in spring and he/she had fed sugar syrup and a pollen supplement or substitute early enough and the weather is warm and settled, a surplus of apple honey many be gathered from a large orchard setting. A lot of if's, but it is possible.

Cucurbits are not that obliging. The male flowers produce a lot of valuable pollen, but the separate female flowers are not very good nectar producers. In fact, bee-keepers who pollinate cucurbits, cucumbers in particular, have to be extremely watchful that the bees do not run out of honey and starve while on cucumbers. They invariably must feed sugar syrup when they remove their colonies from pollinating the crop.

The extreme opposite of a no-nectar pollinating situation is the orange blossom is Florida and California. So much honey is normally made that beekeepers generally receive no compensation for the important pollination their bees provide.

Grapes do not require honey bees for pollination, but the grape blossoms do yield some nectar. If you are fortunate to have access to a large acreage of grapes and all conditions we mentioned above are met, protection from pesticides included, you may be fortunate to have your honey bees collect a small amount of grape nectar.

Remember, all the crops you have mentioned are minor or non-existent nectar producers because of the differences mentioned. Experiment with a few colonies on these crops and have a good time while you are learning.

APPLE POLLINATION

CHAPTER 18

MISCELLANEOUS

CAREOZZO'S HONEYCOMB CONSTRUCTION THEOREM

Exposing the *Myth* that bees purposefully construct hexagonal honey comb cells because it is an efficient shape for storing honey and minimizes the amount of wax required.

My theory is that bees *do not* intellectually and conscientiously set out to construct six-sided (hexagonal) cells in which to store honey and hold eggs and growing larvae.

Bees *do not* instinctively have a natural tendency or impulse that gives them the ability to construct a complex geometric design (a six-sided, six-angled hexagon) into the cells they build for their honey combs. But over the years we continue to see pictures of honey combs and read descriptions such as these...

1) The bees then *mold* this wax into hexagonal (six-sided) cells. A hexagon is a very efficient shape for storing honey. How do bees know that a hexagon and not a circle, triangle, square, rectangle, octagon, or any other figure is the ideal shape in which to store honey? We don't know. (From: *Friendly Bees/Ferocious Bees*...by Kerby)

2) The liquid hardens into wax scales that the bees *form* into hexagonal, or six-sided cells. (From: *Killer Bees*...by Kathleen Davis & Dave Mayers)

3) Workers *make* six-sided cells from wax. (From: *Insects–Eye Witness Explorers*...by Steve Parker)

4) Both sides (of the comb) have geometrically perfect hexagonal cells, a shape which *minimizes* the construction material required. (From: *Animal Life Encyclopedia* Vol. 2...By Grzimek)

In all four of these references and eight other encyclopedias, there is an implication that "Intellect" played an important part in the construction of these hexagonal cells that comprise the honeycomb. Example...The bees *mold* or form or make these six-sided cells *because the hexagon is an efficient shape for storing honey and minimizes the amount of construction material required. That's nonsense.* The solution is so simple I'm surprised that this myth has continued for all these years. First, it must be understood that bees produce heat. A lot of bees produce a lot of heat. In fact on a hot day, bees by fanning their wings, actually cool the hive down by circulating and moving air in, around and out of the hive.

The second thing to understand is that the bees produce honey and beeswax; and beeswax, like the wax of a candle, melts when heated or warmed. *And that is really the solution to how the cells take on their hexagonal (six-sided) shape.*

The bee is actually constructing a *tubular (circular) cell* that fits its body so that the bee can enter the cell to fill it with honey and to nurture the larvae. These tubular (circular) wax cells are stacked one on top of the other until the honey comb is completed.

Now to understand how the transformation takes place from circular to hexagonal, you must visualize or experiment with two balloons.

Take two equal size balloons and press them together. When the two surfaces touch and the balloons are pressed together with a little force, the surface between the balloons becomes a *flat surface*. Now if you take one of those balloons and place it in the center, then circle it with *"Equal size"* balloons, it will take exactly six to encircle it completely. No matter what size circle or balloon you use, *it still only takes six of equal size to go around it completely*. Now put pressure on the six surrounding balloons so that all the pressure is directed to the center one and see what happens...It *automatically forms a six-sided hexagonal geometric shape*. The same principle applies when using tubular (circular) wax cells.

So the bees don't sit around thinking about how many sides they have to construct to form a hexagonal cell. They just make tubular (circular) wax structures to fit their bodies...Then the *slight pressure of the other cells combined with gravity and heat turns them into a mass of hexagonal honeycomb cells.*

And that is my theory...Bees don't construct hexagonal cells. It just happens in a natural way–soft wax, gravity and heat do it. Yet every source over the years has continued to *insinuate or imply that bees, through intelligence and reasoning, came up with a six-sided cell structure because it is an efficient shape for storing honey and minimizes the amount of construction material (wax) needed.*

This is simply not true and encyclopedias should start correcting this error in future editions.

Guy Careozzo
Fountain Valley, CA

Answer - In thinking about your ideas, a question formed in my mind. Several years ago honey bees were taken aboard the Space Shuttle to study their comb building behavior in a weightless environment. If I remember correctly, the honey bees did have a hard time in orienting their comb without a gravity clue, but the individual cells were still hexagonal in shape. How could this be if your theorem is to be considered?

WINTER STORES

Question - With winter weather as severe as it is (Wisconsin), how do you make certain that colonies have enough winter stores?

Answer - This is always a difficult problem, especially as the weather becomes less predictable and seemingly more severe with each passing winter. You're not going to be able to guess the exact amount of winter stores needed, so if we have to be wrong, let's error on the side of giving too much rather than too little and losing the colony.

Start thinking in terms of preparing for winter during the fall honey flow by not placing too many supers on the colony. Additional supers should be added only when the one on the colony is practically full of honey. This will prevent the bees from spreading the incoming nectar throughout more supers than they will fill. This becomes important when you remember that bees may not be able to leave the cluster in search of scattered cells of honey. There have been several reports of colonies starving even though there was honey to be had in the colony.

During the removal of the fall crop, one super completely filled with honey should be left on the colony. In addition to this super, the colony should have from 20 to 30 pounds of available sealed honey in the brood nest. This honey in the brood nest will usually be found along the tops of the brood combs. The honey in the super, plus the honey in the brood nest, should give the colony from 60 to 70 pounds of honey for their winter and early spring consumption.

If the fall flows are uncertain and little fall surplus honey has been gathered, the bees will have to be fed a sugar syrup for their use during the winter. The minimum amount of stores (either honey or sugar syrup) necessary for the winter consumption of the colony is 50 pounds. The beginner may find it convenient to take a pair of scales, similar to ice scales, and weigh the colony of bees. A complete hive with its bees should weigh about 40 to 50 pounds with the top cover removed and any weight over that may be considered as stores. For example: if a colony weight is 75 pounds then it has 35 pounds of honey. To gain the additional 15 pounds necessary for minimum stores, sugar syrup should be fed to the colony. The best feed for winter stores is made by mixing two parts sugar to one part warm water. A gallon of this

mixture increases the weight of the colony about seven pounds so a colony weighing 75 pounds would need two gallons of sugar syrup to bring it up to the minimum stores necessary for successful wintering.

Common practice is to leave a second hive body on each colony, one with sealed stores of honey and pollen being desirable. Record the amount of winter stores left on the colonies in your record-keeping system so you will have a guideline for the next winter.

Langstroth weights:
1 body – top – bottom = 40 lbs.
2nd body = 30 lbs.
1 super = 12 lbs.
Lid, telescoping = 8 1/2 lbs.
Lid, flat tin = 6 3/4 lbs.
Lid, flat wood = 6 lbs.

Question - 1) I want to build a solar wax melter. Is there any advantage of using stainless rather than galvanized for the tray and pan?

2) If I use lumber treated with arsenic for the wax melter box and frame, will the arsenic get into the wax and be harmful to the bees?

3) Will a stainless or a galvanized strainer or pan/tray make honey/wax lighter in color?

4) The old type milk strainer was big at the top and fit a 10 gal. can. Most I remember were galvanized and used a cotton gauze type filter. How could these be modified to strain honey? I have a 20-frame radial extractor. I intend to build a stand to raise it high enough to let the honey run through a strainer into a 19" dia. tank, what size strainer and what mesh screen, would you recommend? Most of my extracting will be in July-Sept., so it should be warm to hot. Kelly's 1996 catalog, p. 42, Strainer Screen, at least part of it is galvanized. Will all or any part of a strainer being galvanized make the honey darker?

5) In *Bee Culture*, page 608 Mailbox, Olive oil treatment. In your opinion could any vegetable oil/vegetable shortening be used for monthly mite treatment throughout the summer months?

6) In *Bee Culture*, page 607 Mailbox, what is Mite Solution? Who handles it? How long will it be before it is available through bee suppliers?

7) In late April through May when you are checking hives and you find a queen cell, can you take the frame and brood with the queen cell and bees that are on it, put it in a nuc box and expect any success in establishing a new colony?

Myron Denny
Stillwater, OK

Answer - 1) A good quality stainless steel would be the optimum choice of any material to come in contact with honey. Honey is slightly acidic and will react with more common materials such as galvanized, iron, copper, aluminum, etc. To optimize the purity and looks of your honey and beeswax recovered from a solar wax melter, stainless is the best choice.

2) The arsenic could only leach into the wax if the wax is in contact with the treated lumber. Your stainless pan should prevent this.

3) A galvanized pan may make the honey/wax darker because of the acidic honey reaction with the galvanized coating.

4) The size mesh of your strainer or strainer cloth will in some measure depend on the temperature and clarity of the honey being strained. If the honey is full of wax debris, it will clog any strainer more quickly than a cleaner honey to begin with. Start with a paint strainer arrangement that starts with bigger mesh in the first stage and progressively gets smaller as debris is removed.

5) It is not approved for use during honey production. But, being the substance it is, it should not cause any harm to the honey produced during the summer.

6) The EPA has started legal action against the maker of the substance because of irregularities in the research, sale, and the advertisement of the product. You won't see it any time soon.

7) Sure can. One of the best ways to start a new colony.

Question - Does beeswax have any health benefits? I eat it with comb honey on bread and have never felt or read that there are benefits.

Is there a hygienic reason why scrap comb honey (not contaminated honey) cannot be fed to honey bees for recycling?

Are swarm queens small enough to pass through a queen excluder, being that she has slimmed down for her flight?

I know the usual method of disinfecting hives, i.e. boil in lye. Would not soaking in Clorox bleach do the same thing?

After Foulbrood disease has been treated with Terramycin and seems to be disease free, are the hive parts still contaminated and will they recontaminate the brood?

The explanation of the wash boarding behavior of bees in the front of the hive has not been convincing: scraping and polishing the hive entrance. I have observed the wash boarding by the bees on the sides of the hive as well. Could it be the act of wax-making, like the chain formation (festooning) inside the hive?

Pollen is collected to feed the bee larvae. Do adult bees eat it also?

Fred Fulton
Montgomery, Alabama

Answer - *Beeswax is not digestible. I suppose because it is indigestible, it acts in a similar fashion as roughage which we are told to eat more of these days. Beeswax does have small amounts of pollen in it which would be digestible.*

If you are confident that your "scrap" comb honey does not contain any disease causing pathogens, use it to feed your bees.

Back in the deep dark recesses of beekeeping research, somebody came up with average dimensions of a queen's thorax. This then allowed standardized queen excluders to be built and designed not to let a queen's larger thorax through. A queen excluder does not work on the principle of abdomen size. Let me give you Jerry's Law #1. "An average is a generality and all generalities are false." Therefore, some queens have thoraxes smaller than "average" and can get through a properly made queen excluder.

Although Clorox is a good disinfectant, we cannot recommend it for decontamination of foulbrood-infested equipment. Burning the frames and scorching the hive body interiors is the only universally recognized AFB control measure. Even the more modern Ethylene oxide fumigation chambers are not recommended any longer because of the risk of carcinogenic residue being left on the bee equipment. Boiling lye water is messy, warps frames and may not be 100% effective. More recent research shows promising results with gamma irradiation, but this decontamination method is still in its developmental stages and is quite expensive even if one is located near a gamma irradiation facility.

Terramycin, like any antibiotic (antibiotic means, anti-against biotic-life) kills any organism not in the protective spore stage. AFB is survival oriented. It survives by forming a resting spore stage that is impervious to many chemicals, boiling, high heat and aging. When conditions are favorable again, it comes to life. The answer is yes, if there are spores, they are waiting to recontaminate the brood.

Your guess on wash boarding behavior is as good as any except wax generally isn't secreted at this time.

Yes, adult bees do feed on pollen. Bee larvae feed on worker jelly secreted by adult bees after ingesting pollen. Let me quote from page 322 of the new edition of The Hive and the Honey Bee: "The health of the honey bee colony is dependent on pollen as it is upon honey. Pollen is virtually the sole source of proteins, fatty substances, minerals, and vitamins that are necessary during the production of larval food and for the development of newly emerged bees. A colony cannot rear brood if it does not have pollen. Older bees can rear brood without consuming pollen, but they do this at the expense of their own bodies and the amount of food produced is rather small."

Regarding larval nutrition, let me quote from page 224 of the new edition of The Hive and the Honey Bee: Worker bee larvae are fed a glandular secretion produced by the mandibular and hypopharyngeal glands of adult nurse bees, normally during the first 2 weeks of their adult life. The secretions of the hypopharyngeal glands are clear, containing mostly protein, while those secretions from the mandibular glands are white, containing mostly lipid components. Worker larvae are fed a so-called "worker jelly" which consists of 20-40% white component for the first 2 days of larvae life. On the third day the white component ceases to be fed and for the last 2 days of larval development only the clear secretion from the hypopharyngeal gland, mixed with honey and pollen, is provided. Pollen supplies little of the nitrogen requirement of larvae and may be a contaminant from honey stomach fluids (Haydak 1970). This diet is often referred to as "modified worker jelly". Drone larvae are probably fed a similar diet, but they receive more food than worker larvae.

LOTS OF QUEEN CELLS

QUESTIONS ABOUT AFRICANIZED HONEY BEES

Question - There are many, many questions on the minds of beekeepers about Africanized Honey Bees, such as, how many hundreds of swarms are crossing the border each year? What percentage of swarms are captured at pheromone baited trap lines compared to the number that fly right past them? In warm areas are these colonies able to establish themselves within shrubs or tree branches when hollow trees or other protected sites are unavailable? What is the proximity of migrating swarms to California and Arizona borders? For each colony discovered and destroyed, how many more remain undiscovered? I could go on and on and many readers would have dozens of other questions.

Robert Weast
Johnston, Iowa

Answer - *Courtesy of Dr. Frank A. Eischen, assistant research scientist, Texas A & M University, Weslaco, TX.*

Good questions all. First off, because the geography involved is on a large scale, precise answers are not always possible. The USDA-ARS has pheromone baited swarm traps along the Lower Rio Grande River starting at Brownsville and running upriver for about 100 miles. Cluster of four traps are set out every five miles. Last year they caught 272 swarms, about 60% of these were Africanized. We know from work done by Dr. Bill Rubink (USDA-ARS) in Mexico that the success of these traps depend on the number of good nest sites in the surrounding area. Where there are many natural cavities, the traps success is not good. In areas with few natural nest sites, their success improves. Still, we don't really know. Some experts have suggested that 10% or fewer of the swarms in the area of the traps are captured. One last word on this. The trap lines were designed as a monitoring device. There has been no attempt to use these as a control technique. The number of traps and the time required to service them regularly would be prohibitively expensive. Some would further argue that it would also be doomed to failure even if we had the money.

It's frequently been reported that Africanized honey bees nest in odd places, eg., rubber tires, old water heater tanks, electrical boxes, etc. Perhaps this is because they are less demanding that European bees when choosing a nest site. Our experience in the Lower Rio Grande Valley confirms this. We sometimes find relatively exposed nests, eg. suspended from a tree limb or nests built in thick tangle of sugar cane. Dr. Chip Taylor reports that he has found nests built in thick clumps of brush in Mexico. These can be good defensive positions from the bee's point of view. Brush in this area tends to be thorny and nearly impenetrable. Any animal or person attempting to get at the nest not only find the going pretty tough, but each motion tends to telegraph its action to the bee nest. The vibrations also do a fine job of irritating the colony. The large size and dark comb of some of these nests suggests that they have survived for a relatively long time. Our winters in South Texas tend to be short and quite mild, so climactic condition should allow some of these nests to survive, if they have enough honey. Counterbalancing this observation is our finding that many Africanized colonies have remarkably little honey. Relatively short periods of inclement weather could be devastating for these colonies. An exposed nest can then become a deciding survival factor.

Africanized honey bees have been migrating up the western side of Mexico a bit slower than those on the eastern side. Dr. Jorge Julian Gonzales, Subdirector, National Program for the Control of Africanized Honey Bees, reports that the last northernmost find is in Ciudad Obregon, Sonora. That occurred five months ago. He suspects that they are now in Hermosillo, which is about 120 miles south of the Arizona border.

I think it's safe to say that the vast majority of Africanized honey bee nests go undetected. Thousands of square miles of South Texas range land are unmonitored. Swarm trap lines at the leading edge of Africanization serve to detect new advances only. In the Lower Rio Grande Valley it's likely that a great many Africanized colonies have been killed, perhaps thousands. This area might be thought of as urbanized farm land. A swarm of bees here is likely to be seen by someone and reported. City, county, state, and federal agencies have been involved in killing Africanized swarms and colonies. It's probable that a significant portion of the Africanized population has been killed. However, there is probably a net movement of Africanized colonies into Texas from Mexico, and undetected U.S. colonies continue to swarm.

Dr. Frank Eischen
Texas A & M University
Weslaco, Texas

Question - It was brought to my attention while talking to a friend about putting some hives on his property that if the neighbors are spraying their flowers, dandelions, etc. (things that the bees will be visiting), won't the bees be bringing those insecticides into the hive and virtually into the honey?

Here in Boise, Idaho the temperature during this winter is ranging from 30 degrees up to 48 degrees. So far it has been a mild early winter. I have noticed that the bees in my hive have been coming out when the sun is out and it's warmer. My question is: Can I feed them sugar water to help them through the winter? There may be some days that the jars freeze, but I don't have that many hives that I couldn't resolve that problem quickly.

Mark Gosswiller
Boise, Idaho

Answer - Honey bees either accidentally or purposefully bring in a variety of things from their environment–pesticides, herbicides, fertilizers, paint, tar, pollen, etc. If you do not live in a pristine wilderness, honey and pollen will be "contaminated" (that might be too strong a word) with man-made and biological excess and residues. The beekeeper only notices a problem when one of the substances kills large numbers of bees in the hive that he/she can see.

The bottom line here is honey, pollen and beeswax were never "pure" substances. However, in most cases the contaminants (either natural or man-made) are harmless to man and bee. Pesticides are a different story, however, and if either your neighbor or nearby farmers spray heavily on blooming flowers, you may have to consider moving colonies away from the problem.

I'm glad your winter has been mild so far. It makes it easier on bees and people.

A colony of honey bees generally will start to cluster when the temperature reaches 47°F. But with the sun shining, small micro-climates may exist which allow bees to fly out and about on these days.

It is a temptation to want to help your bees as much as possible, supplementary food included. If by lifting, tilting, visually inspecting your colonies, they have an average of 50-60 lbs. of honey stored, that is plenty to get them through the winter and spring without additional help. Honey is the best food for them at this time.

If you find less than 50 lbs. of stored honey, you may want to immediately start feeding or delay awhile, depending on the actual amount stored.

An outside feeder is generally a bad choice for emergency supplementary feeding because temperatures generally are too low to allow enough bees to consume enough sugar syrup to make a real difference. An internal feeder, either a bucket type of division board feeder, can allow the bee cluster to access the feed more easily and consume a lot more–enough to make a difference. Candy boards or dry sugar are two other options if the weather is too cold to feed syrup.

First, determine your bees' real needs. Then decide how to fill that need if there is one.

Question - I started a hive of bees in the spring of 1992. I ordered all of my equipment from Dadant in Durham, California. I have also ordered several books on bees from Dadant. However, I have a problem and since there are not many beekeepers up here, I wonder if you could answer a couple of questions for me.

I live in Volcano, California at an elevation of four thousand feet. None of the books that I have read, thus far, give much information on raising bees at this elevation. I have tried to compare what my bees are doing to the books, figuring that my bees would be about a month behind what the books say should be happening at any given time of the month, but I cannot get a good match. Do you know of any books that deal with raising bees at higher elevations in the timber as we are?

Secondly, last fall after we harvested the honey we noticed two things that were different from what the books said and from store-bought honey. First, we only got about 26 pounds of honey from the one hive. The books said that there should have been about 50 pounds. We think the reason was that there were not as many flowering plants that year because of the drought. Second, the honey that we did get was very sticky, so much so that we have to heat it up in a pan of boiling water before using it because it will not spread or come out of the jar very well. Are these conditions normal for honey from the mountains?

This year we are putting in a second hive since the bees appear to have survived the winter very well, even though we have had more snow this year than we have had in the past six years of the drought.

I hope that you can direct me to some informational materials that might help me with my problems and questions. Eventually we hope to have 25 to 50 hives in this area.

Michael Hammil
Volcano, CA

Answer - No, I don't know of any specific publication that addresses your particular region for beekeeping questions. I, like you, would have thought that elevation relates to temperature and thus you could extrapolate from there. Apparently not for your situation. I've included a list of your California "powers that be" and a list of California regional associations. I'm sure that answers are available from some of these. You can't be the only beekeeper who has experienced these conditions in the whole state.

Your honey just sounds like very low moisture honey. Most honey has moisture levels around 18%. But in selected circumstances i.e. the desert southwest or in some drought areas, honey moisture can get down to the 12-13% area which makes it, of course, less liquid. There is no honey in the U.S. that I know of that is naturally such as you describe, unless it has been reheated so many times that the moisture has been forced out. If your honey has crystallized it will, of course, be thicker and will need to be heated to return it to liquid form if you prefer liquid honey.

SPRING SPLITS, ETC.

Question - 1) When you make a split, placing it on the top of the mother hive, using a double screen, how do you get your split to accept your new queen which is still in her cage? Won't the pheromone scent from their mother queen down below induce them to think they still have their queen and won't they kill the new queen in her cage?

2) You have advised us we can use partially filled frames the following spring for feed. How do you keep the moth out of these frames until winter weather? Can we use the normal Para Crystals treatment? Will that odor be gone enough by next spring? What about partial frames that aren't capped, how do we keep them from crystallizing or will they still feed on them? Can we use the crystallized honey as feed, too?

3) We want good public relations, right? When a person needs to be told how to reliquefy honey, then surely he should also be advised to remove the lid of the container when heating. Why don't these label people include this in their reliquefy-ing labels? When someone heats honey and a jar bursts all over her kitchen, you can be sure that this person will never, ever buy honey again.

John Anderson
Arcola, IL

Answer - Question #1 - Honey bees pheromones are important for all aspect of colony organization. While some pheromones are fairly volatile (diffused quickly in the air), others such as those which suppress queen rearing and let the colony know that it has a queen are disseminated primarily by oral contact between bees.

You may have read about or seen the ever changing group of "attendant" bees that groom (lick) the queen. Well, they are not really grooming her like one thinks of grooming a dog. They are licking off various chemicals that the queen is secreting. These chemicals tell these bees information about the status of the queen. These bees then share the chemicals with other bees in the hive in their food-sharing behavior called trophallaxis. So, as the bees greet each other and share a bit of honey or nectar that also has these queen chemicals in it, the information about the queen is passed continuously through the colony. When a queen is removed or a split is made separating some of the hive from the queen, it takes about 8 to 12 hours before the queen chemicals run out of the bees passing them and the food within the colony. At this point the colony then starts trying to do what it can to rear a replace-ment queen or accept a new queen that the beekeeper has introduced. After this time, if a good caged queen is put into the split, the bees within the colony will be very anx-ious to start the licking/grooming behavior through the screen on the cage. And about 8 to 12 hours later, this new queen's chemicals have been distributed through-out the colony and she is the queen now in charge and the one accepted as their new mother of the colony. It's always best not to rush the introduction phase for a new queen to be sure her pheromones have been thoroughly distributed throughout the colony.

Question #2 - The easiest way to keep combs of honey available for feed for any of your colonies is to group them together on designated hives. If you leave sufficient stores on all your colonies and then put two or three supers additional on one colony, chances of wax moth infestation are reduced and you don't have to use any chemicals. Then, if you need this honey to feed your colonies in that yard, they are available right there. You don't have to transport them from somewhere else. Some honey may crystallize regardless where it is stored. In most, but not all cases, it can be used for bees needing an extra boost in spring.

Question #3 - You're right. Some of this burden of providing product information to the consumer is the producer's responsibility. We can't expect or assume that the distributor or consumer knows everything about honey. We have to educate them. If we don't, I don't think our competitors for shelf space in the store are going to do it for us.

Question - Do bees see in the dark??? The inside of a hive is dark, but yet bees find each empty comb in which to deposit nectar. The queen finds empty cells in which to lay eggs and the bees know when to cap each cell brood of honey.

Do bees see in the dark??? I have tried to answer this one myself. I placed a dish of 2:1 sugar water near some hives late afternoon when no nectar flow was on. The bees worked it well till dusk, but slowed up at dark and then stopped. There was more sugar water, but the ants found it. Early next morning before sun up I checked and watched. Sure enough, the bees were starting out soon after I could see just where the dish was.

I don't think they see in the dark or they would have worked the sugar water till empty all night.

Charles M. Starkes
Virginia Beach, VA

Answer - Mr. Starkes, you are absolutely right that honey bees do not "see" as we do in the dark, and your well reasoned experiment proved it.

Honey bees have two compound eyes and three simple eyes on the top of their head. These are used to navigate during daylight hours. But, they are useless at night or within the confines of the hive.

Honey bees, like many other animals, including other insects, mammals, birds, fish, etc., transfer a great deal of information using specific chemicals that are produced for a chemical language. Let me quote from Chap. 9, page 373 of the new edition of The Hive and the Honey Bee, "Although honey bees transmit some information with acoustical (sound) and visual signals, most of their known messages are generated with chemical cues. It is now possible to associate many of characteristic behaviors of queens, workers, and drones with specific compounds that have been identified as key elements in the honey bee's chemical 'language'. The availability of these signaling agents explains how it is possible to rapidly galvanize the great resources of the colonial members in order to exhibit the specialized insect societies.

In essence, insects like the honey bee have evolved a chemiosocial lifestyle that guarantees that the collective resources of the colony are available to be utilized whenever required."

Like our verbal language, honey bees use certain chemicals that have meaning all their own. Just like when you smell chocolate chip cookies baking, you don't have to see them to know what is being cooked and what is in store for you later when they are done!

Question - Recently, while inspecting a nuc hive I noticed something different. There were a few bees that had a set of scale like formations on the extreme of their front legs. These scales were attached to the tarsal brush section of each first leg. I inspected the literature for some clues on a disorder like this without any positive findings. I have included a sample bee with these scales on its legs for your use in identification. Do you have any information on anything similar to this?

Felix Torres

Arecibo, Puerto Rico

Answer - Thank you for your interesting question and sample of the honey bee with the "things" attached to its legs.

When I first looked at them, I had no idea what they were. I had never seen anything quite like them and did not know if they were animal, vegetable or mineral. Truthfully, I was a little bit nervous that I was going to be involved in your discovery of some devastating new honey bee parasite that was going to decimate the beekeeping industry.

I then took one of these very, very small winged shaped things and looked at it under a microscope. I could tell easily then that it was some kind of <u>plant</u> structure. But, what kind of plant has this capability?

It was time to call in a better mind than mine. I called Dr. Shimanuki at the USDA Bee Research Lab. in Beltsville, MD. Probably some time during my 30 seconds of explanation and description of what I saw, I'm sure he knew what it was, but he was kind enough to wait until I had finished. He told me it was "pollinia", usually from the Milkweed plant. Now I was sparked to find out more about Milkweed pollinia. Here's what I found out.

Milkweed and some Orchids, have a somewhat peculiar pollination mechanism. The pollen is not presented in loose dusty single grains, but stuck together in two separate groups (a pair) which are attached on a short stalk at a junction forming a wedge-shaped clip. Picture a wishbone. The two long projections on the wishbone would hold the mass of pollen and the center structure where these meet is the wedge-shaped clip which snaps and attaches to the bee. The bee then would take this pollen packet to another flower and perhaps act as the mechanism for cross pollination.

If there are but a few of these pollinia attached to the honey bee, the bee can usually still fly and maneuver until the pollinia can be removed by grooming. If there are many pollinia stuck on the bee, they may restrict flight and even crawling because of the mass of the pollinia hanging on them.

This situation is fairly rare, but not rare enough that beekeepers haven't had similar problems with it in the past. It's not a fungus, bacteria, mold or parasite. It's just the milkweed pollination system that the bees are participating in, sometimes to their detriment.

Thanks for the question of "What is this?" I know I learned something, I hope you did, too.

Question - I have a couple of questions I am wondering if you could answer for me. First, when feeding bees is there anything you can put into the sugar syrup to keep it from spoiling? After setting in the sun, even in cooler temperatures, a green scum will begin to form in the jars. Secondly, as a novice beekeeper, I sometimes dream about doing this for a living. In spite of some of the market and mite problems, and some commercial beekeepers going out of business, I would like to know how one can become a commercial beekeeper? What information is available to explain what it takes in terms of investment, time, equipment, number of colonies, etc.? Any information you could give would be appreciated. Thank you.

Jerry Etringer
Cedar Rapids, IA

Answer - You have a couple of real good questions. I hope that I will be able to give you some answers that you can use.

Theoretically, the beekeeper would only give enough sugar syrup that the bees can consume in 24 to 48 hours. Yeah right! I give my bees a gallon or more at a time and rarely in early spring do they finish it up in less than a week or ten days. I used to get that mold or fungus stuff in my feeders also. I don't know if it affected the sugar syrup to the extent that it was harmful to the bees, but just by looking at a moldy fungus decorated bucket, I don't think it helped them either. I learned from a beekeeper in California that if you put approximately one teaspoonful of regular household bleach per gallon of syrup that the growth of mold and fungus will stop in your feeders. The bees don't seem to mind and there is some anecdotal information that this practice cuts down on the increase of chalkbrood. This has not been substantiated by formal research, but it would make some sense. Give it a try.

I sometimes dream about being a commercial beekeeper, too. It's a good dream to have. Now do you want to produce honey only, pollen only, devote your time to providing pollination or do you want to be a queen producer or package bee producer? Do you want to pack your own honey and other's honey under contract as a private label packer? How about producing a clean pesticide-free pollen product? Either dried or dried and packed or maybe a blend of pollen and honey–maybe capsules?

What about feeding your bees a lot in spring, raising some queens and producing nucs or packages? How about taking your bees to a grower's orchard and being paid for pollination, then still produce a honey crop, bottle it yourself and sell it to grocery stores?

I could go on and on. The opportunities are almost endless. But to be successful, you must be a self-starting hard worker who knows bees. A self-starting hard working salesman, a self-starting hard-working marketer, merchandiser, public relations person etc., etc. The problem here is that most people cannot do all these things well. Then, instead of hiring out the things they don't do well to someone else who does, they try to muddle through and never reach the business potential they thought they could or should. Realize your strengths and weaknesses and don't let pride take you the opposite way to go. My suggestion is to visit as many sideline and commercial beekeepers that you can and ask them as many questions as you can. Information like this is priceless. Then, plan your work and work your plan. Be sure to start small and build up gradually! Good Luck!

CLASSROOM COMMENT

Jerry, your answer to Jerry Etringer may stir up some fuss since chlorine bleach is probably not an approved chemical for use in the hive. I have no specific objection on those grounds, but I believe using a teaspoonful per gallon is way overdoing it. Chlorine, even as sold diluted in household bleach, is very potent. If I had to make an educated guess at the maximum dose you should try, I would go with no more than 3 parts per million, which would only be a few drops of 5.25% chlorine household bleach per gallon. How would you know if you achieved the right concentration? Easy, get an inexpensive chlorine test kit as sold for spa or swimming pool testing or home water supply testing. You could also do some math to see how many milliliters of bleach would be needed for a gallon of water. What harm can you do by over dosing? Bees, like other creatures have beneficial digestive bacteria in their guts to help digest their food. Your solution will greatly reduce those "good" germs. Have you ever swallowed too strongly chlorinated swimming pool water? It gives you a belly ache and cramps and will "run" you to the can big time! I would not wish that on a bee. Despite some flack you might catch about chemical abuse in the hive, the right amount of chlorine might be a good sanitary practice for the bees' sake.

Thanks for the good info, Vernon.

Jerry

Question - What are the pro's and con's of using almond extract (Benzaldehyde) on fume boards instead of Bee-Go.

Kay Almond
Pocatello, Idaho

Answer - Until about three or four years ago a popular alternative to using Bee-Go was a product called Benzaldehyde. Benzaldehyde is artificial oil of almond. It was widely available from all beekeeping supply distributors. The product has been removed from the list of approved honey bee repellents because of possible health concerns for the users of benzaldehyde. So, at this time the only approved alternative is Bee-Go.

Question - Will fermented honey harm bees if fed to them in a hive? If a brood chamber is almost full of nectar/honey, with no pollen stored, will the workers move the nectar up to the supers so that there will be room for pollen and the newly installed queen can lay? Is it a fact that workers do move nectar as needed?

In an extremely dry summer when no blooms/pollen are available, will the queen stop laying? Will the workers kill the queen? I have lost queens several times this summer.

Fred Fulton
Montgomery, AL

Answer - Will fermented honey harm bees if they consume it? Sure it will just like it harms any other creature. Alcohol always damages tissue. Nerve tissue is always noticeably and measurably affected. I've always thought it interesting that the excrement of yeast has gained so much attention in our world. Don't feed honey bees or anything alcohol! We've heard and read that fermented honey will cause dysentery in honey bees, especially if fed during the winter.

Honey, once deposited in a cell and capped, usually stays in that cell until consumed. Fresh nectar, on the other hand, is spread about the hive to quicken moisture evaporation. Then, as the honey "ripens", full cells are capped.

Bees cannot raise brood without pollen for protein. When the pollen source is used up, all egg laying stops. If it stops for a long enough period of time, the workers become alerted that the colony is not maintaining population and will replace the queen, reasoning that she has failed. The workers are reacting to no brood to feed and a decrease in the pheromone which tells them all is right in the colony.

Question - 1) Is there any medical value to beeswax?

2) Isn't it possible to feed swarms too much sugar syrup or scrap honey? Early in the life of a new colony it seems that workers may pack the hive cells with these liquids leaving the queen with no empty cells to lay.

3) In a back issue of *ABJ* I read that chloroform in a hot bee smoker tames the most hostile colony. Chloroform is not available to me in my local pharmacies or farm supply stores. Solution?

4) Again in *ABJ*, I read that there is a significant queen loss by queen breeders due to predation of queens during the mating flight by birds. What is the prevention?

5) Do bees recycle discarded wax from uncapped cells? Honey cappings are new white wax, while other cappings are always off color.

6) Drone cells on one side of a brood frame usually have corresponding drone cells on the exact location on the reverse side. Why?

7) In crushing drone cell cappings with my hive tool, I have observed at a later date worker bees helping some of their trapped matured brothers free themselves from the damaged cells by eating away at the cell edges. Isn't this unusual? I thought worker bees ate the damaged brood.

Fred Fulton
Montgomery, AL

Answer - 1) Beeswax is not digestible. Therefore, the only possible value it may have is that of roughage in one's diet.

2) It would only be too much if the beekeeper was not observant enough to recognize the need for more storage space i.e. hive body with frames and foundation or drawn comb or if he or she kept feeding, causing the problem you mention.

3) Chloroform is a powerful anesthetic. It is not used in any degree in the medical profession any longer due to its side effects. I am only guessing, but I suppose it would cause a stupor in the workers or possibly kill them. Let's think ahead a little though. What would this chemical do to the queen? Would it kill her, stop her from laying, shorten her life span, etc., etc.? What would it do to the uncapped brood? What would it do to the honey? Would it be absorbed and make the honey unfit to eat for the bees or man? What would it do to the wax? Would the bees then abscond? These are a few things we need to think about first before introducing any chemical into the hive. Sure, it may do one thing well but do damage, real damage, in other respects that would make the hive useless.

Only use products which are government approved for use in a bee hive.

4) What is significant for a queen breeder? I called a few and they told me that only between 1% and 3% of virgin queens do not return from mating flights. You never want to lose any, but considering the precarious nature of honey bee queen mating, it's really not too bad.

5) Bees move wax all around the hive. It's quite energy intensive to produce wax, so they use it as wisely as possible. Honey cappings are pure beeswax because the honey must be tightly sealed to prevent it from gaining moisture from the air and pure beeswax will do this.

On the other hand, sealed brood needs oxygen coming in and carbon dioxide leaving the cell to live. These cappings are a combination of beeswax, old larval skins, and propolis which makes porous cappings which are darker in color than pure beeswax cappings over honey cells.

6) Structurally, the master engineers that bees are, they have chosen this as the most secure method to build these cells.

This brings up another question, why do honey bees build hexagonal cells instead of square or octagonal cells for example? Engineers tell us the hexagon provides the strongest structure which make the most efficient use of space. Okay, that's fine, but who told the bees this? Scientists will tell you it's all a matter of evolution, instinct, and pheromones. Honey bees are preprogrammed minicomputers if you will. That's an easy answer, but it doesn't explain the complex mechanics of honey bee behavior which trigger this comb building instinct or how the bees measure and construct such uniform architectural structures. There remains much we do not understand about honey bees.

For more information on this fascinating subject, we suggest you read H.R. Hepburn's <u>Honey bees and Wax</u> published by Springer-Verlag. Check with your library. It may be available through the Interlibrary Loan System.

A HIVE FULL OF BEES

Question - 1) The bees leaving the hives on foraging trips always seem to quickly brush their antennae with their legs before taking off. Are they cleaning off hive dust, are they orienting themselves as to their location, or do the antennae have special sensors for wind change or the vibrations of other bees' wings that are groomed by that action? Do the legs show a special comb shape that corresponds to the antennae, and, if so, can you find electron microscope or other large-magnification images of the leg-antenna interface to show us?

2) Almost all of the bee traffic in and out of the hive this spring has been on one side of the entrance. Yesterday a bee appeared at the opposite side of the entrance with a very small bit of something black and almost too small to be seen in her jaws. As she was preparing to alight, she dropped it and then picked it up and flew off. One or two seconds after she emerged, another bee with another very, very

tiny black something came out of the hive from the same location, followed by five or six bees who all appeared to be part of the same team, each with her very minute burden of whatever they were ridding the hive of. The question is, what sort of bee activities are done in teams? Does the urge to act in that particular way seem to appear spontaneously in individual bees who then find each other or is there evidence that some bees take it upon themselves to become team leaders who then recruit and motivate other bees to carry out some needed activity? The group observed gave the impression that they were being led.

Pete Caruso
Galesburg, IL

Answer - Thanks for your questions. Watching the entrance to a hive of honey bees on a warm, sunny day is a lesson in itself. And, if one is paying attention, leads to questions. You are doing both which is half the fun of beekeeping.

1) The antennae of honey bees are very important sensory organs. Antennae are particularly receptive and responsive to touch and odor. As an analogy, just think what it would be like to go around all the time with a stuffy nose and gloves on. You could not smell very well or feel very well with your finger tips. Well, that's what it would be like if the honey bee did not have a means to clean her antennae. You can't smell or feel as well with dirty antennae as with clean antennae. Honey bees have been blessed with a structure on their forelegs called – surprise, surprise, the Antennae Cleaner.

For a little honey bee anatomy and explanation of this structure, I have borrowed the next passage and diagrams from pages 132-135 of the 1992 edition of The Hive and the Honey Bee.

The Antenna Cleaner

"The structures used by the bee for cleaning its antennae are situated on the inner margins of the forelegs just beyond the tibiotarsal joints (I). Each antennae cleaner consists of a deep semicircular notch on the basal part of the long basitarsus (A), and of a small clasp-like lobe (J) that projects over the notch from the end of the tibia. The margin of the notch is fringed with a comb-like row of small spines. The clasp is a flattened appendage, tapering to a point and provided with a small lobule (K) on its anterior surface; it is flexible at its base but has no muscles. As this gadget is used by the bee, the open tarsal notch is first placed against the tibia (B), the antenna is brought against the tibial clasp which resists the pressure because of the small stop-point (I) behind its base. The antenna, thus held in the notch of the clasp, is now drawn upward between the comb of the notch and the scraping edge of the clasp. The antenna cleaner is present in the queen and the drone as well as in the worker."

In other words, the bee puts its antenna in the notch and pulls it through. The bristles in the notch clean it off. Pretty cool don't you think?

2) The whole hive is a team with different groups of individuals mostly based on age, doing different tasks such as housecleaning, foraging, guarding, etc., etc.

These activities may change, depending on the needs of the colony, but for the most part are genetically locked in, based on the age of the individual bee. When you have many individuals of approximately the same age you may see, as you did, many bees doing the same thing. You saw several bees bring out debris of some sort. They were house cleaning because they were at the age 1 to 25 days, where their age, environment and genotype dictates they do this.

The tasks required within the hive, while appearing to have some organized "teams", really are only organized genetically by individual.

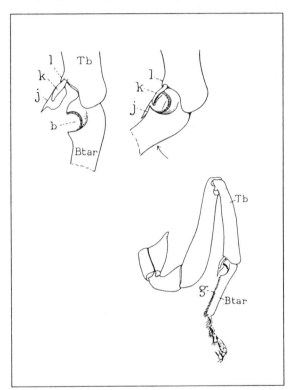

ANTENNA CLEANER

Question - Winter will be here in Michigan pretty soon. Last year I had two hives which I surrounded, leaving some exposure, with bales of hay for wintering. We had deep snow and zero temperatures. On warmer days in the low 30s, I discovered large numbers of bee bodies at the entrance of the hive and sometimes stretching out 60 feet from the entrance. I brought several inside to examine and found the warmth caused them to revive. I don't want the same thing to happen this winter because it seems hundreds of bees may have died needlessly. What was I doing wrong?

Douglas Whitman
Oakland, MI

Answer - *What you saw is not unusual at all. When a colony of bees has been confined for several weeks, there is snow on the ground, the temperature has been cold then warms up a bit, with the sun shining, many bees may exit the hive never to return because it is still too cold outside to sustain the insects' need for warmth. These may be old, sick, diseased or healthy honey bees, but they have chosen to leave. A mistake was made by these bees in judging the actual outdoor temperature and they die. Why do they do this? This brings up the interesting phenomenon of micro-climates. Have you ever been outside on a cold winter but sunny day and been amazed at seeing heat waves radiating off of the road, tree, car, etc.? Well, even though it is generally cold outside, certain structures will absorb sunlight and get very warm even though the air is cold. The ground, bee hives and bales of hay around your hives could have been relatively warm with the sun shining, giving these bees a false sense of their environment. They made a mistake because even though it may have been warm in their immediate surroundings, it wasn't generally. They left, chilled quickly and fell to the ground. Some beekeepers also claim that the brightness of the sun shining off the snow disorientates the bees aggravating the problem even more.*

As you noticed, they can be revived if warmed sufficiently. I suppose you could pick up as many as possible and revive as many as would respond and re-insert them into the hives. I've always preferred to let nature take course. Let it do so this winter also.

Question - My hives are only a few feet apart from each other and I was wondering if this would aid the mites in spreading from hive to hive. Should I put a little more distance between my hives?

I wonder if a dog could be trained to chase and "tree" swarms? A dog that could do that would certainly be helpful to the beekeeper.

Colby McMullen
Williamsville, VA

Answer - *Mites are everywhere and at this point it is of little use to try separating colonies by miles in remote locations.*

A dog's sensitive nose could most likely be trained to locate existing colonies and swarms. With all of the odors emanating from these sources, there is no doubt in my mind that with proper training a dog could do this. Dogs have already been trained to smell foulbrood, just by sniffing at the hive entrance.

Question - 1) Is raw maple sap O.K. to give bees for spring buildup?

2) I've heard that you can get rid of wasps and hornets by mixing Sevin dust with bologna or hot dogs. You place it in a bait box which is dog, etc. safe and the wasps take it back to their nests. I see honey bees frequenting my trash pile and wonder if the bees would also take the tainted meat. Can you tell me?

3) Will my nighttime "bug zapper" kill wax moths?
4) What is the record for the largest comb?

David Ott
Napanoch, NY

Answer - 1) Raw maple sap is probably all right to feed in spring as long as the bees can get out to void. Maple sap is relatively low in sugar in comparison to other sources and it is full of many undigestible substances. It is O.K. but probably has more value turned into maple syrup at $40.00 a gallon than used as bee food.

2) Bees, not being carnivores, would not be after the meat itself. If you look at the ingredients on a package of luncheon meat or bologna, many times you will see some type of sugar high up on the list. With nothing else to forage on, I suppose that honey bees being the opportunists they are, would investigate this concoction. I'm skeptical that they would ingest any pesticide though.

3) "Bug Zappers" may lure in some wax moths, but probably few since these cue in on pheromone trails more than they do various frequencies of light.

As a side note several years ago a study was done to see what kinds of insects were attracted to bug lights and how many were killed. The results were that something like 80% of the insects killed were beneficial insects and the remaining 20% were nuisance insects.

4) I have no idea. I've seen pictures of single combs that were three feet by six feet across which seems like a record to me. In India and Asia, the giant honey bee (Apis dorsata) may at times construct single combs on trees or cliffs which are quite large by Apis mellifera standards.

Question - I've heard from a beekeeper friend in Canada that many commercial bee-keepers experienced tremendous losses of colonies this past winter even with indoor wintering. They are blaming the losses on the kind of sugar used as supplemental food to get them ready to go into winter. Could it be the same problem that caused loss of colonies coming into this year here in the U.S.?

Bob Roberts
Minneapolis, MN

Answer - Colonies were lost in high numbers in both the U.S. and Canada – that can't be denied. Whether the two are related to the same primary causes is debatable. There are just too many variables. The cause of most bee deaths this past winter is a combination or varroa mites and tracheal mites and what they do in the north to goof up the ability of the cluster to thermoregulate the unit. In other words, mites make it hard for the bees to generate and maintain the proper temperature of the cluster in extremely cold extended conditions. In many parts of North America the winter was colder and longer than usual and the bees starved to death even with plenty of honey because then couldn't move to new nearby honey stores. This happened across the upper half of North America.

The specific problem you mention in Canada, as I understand it at this writing (early June), is that "bad" feed was used to feed the bees before going into winter. High fructose corn syrup is used and, if of the proper type, is an excellent feed for bees. High fructose corn syrup (HFCS) is produced from starch, usually corn starch, by converting starch to glucose and then some of the glucose to fructose. There are two ways to do this, an acid conversion method and an enzyme method. Both kinds are fine for human and animal consumption, but only the enzyme method is suitable for bee food. If HFCS produced by the acid method is used to feed bees, it will kill them over time. Also HFCS produced by the enzyme method can be "off spec" meaning it doesn't meet the quality standards for human food production. Sometimes this "off spec" syrup is sold at a discount to use in animal feeds which for a cow or pig is fine. Honey bees are not as tolerant of impurities in "off spec" HFCS, especially in winter when they cannot get out to void. At this time the trail of a cause for some of the Canadian losses points to beekeepers using an unsuitable HFCS for their winter feed.

Question - The August issue of the ABJ had an article by Douglas W. Rawley describing the Stewarton Hive. I have been unable to locate Mr. Rawley, so I was hoping you could tell me how I could get details on this and other ancient bee related items. I have access to the internet and will research there if I know where to start.

I am overwhelmed by all the different bees listed in the bee journals. How does one determine the best choice for their conditions and which have been selected for positive traits. Is there a list describing each strain as you would have for other breeds of animals?

Answer - *My old Alma Mater OSU/ATI has a great beekeeping museum full of old and unique beekeeping paraphernalia in Wooster, OH. Mr. David Heilman is the "Curator" of the facility and is very knowledgeable on older beekeeping equipment. He can be reached at Ph. 330-263-3584 or fax at 330-262-2720.*

The Hive and the Honey Bee describes very well many of the races of bees. Get a copy for this and all of your other beekeeping questions. Another interesting book is In Search of the Best Strains of Bees by Brother Adam.

Question - I've heard that Africanized bee swarms can "take over" colonies of honey bees of European ancestry. How can this happen without tremendous fighting and loss of bees on both sides?

David Anderson
Bulverde, TX

Answer - *The way I understand how a "Africanized Swarm" takes over an established European colony is this: Somehow, probably pheromone levels, a small Africanized swarm with a queen identifies a colony which is preparing to swarm* *itself. The queen is older and may be a candidate for replacement and/or there are queen cells prepared for her replacement. Under these conditions the mini swarm with a queen will replace the European queen with the failing or suppressed pheromones pretty easily.*

Question - I had six hives, but lost one late last fall. One of the five I still have is depositing stuff at the front of the hive. I am sending a sample enclosed.

At first I thought it was wax trimmings. Putting some in water proved it not to be wax granules. I have decided it must be larval remains. Do you have any idea what might be wrong? None of the other hives are doing this. There does not seem to be any more dead bees around it than any of the others.

Roy Goodman
Hillsboro, TX

Answer - *Thank you for the packet of hive debris that you noticed your bees taking out.*

I took a close look at it and showed it to some people here and then put it in some hot water and most of it dissolved. As you note, if it were primarily beeswax, it would float, but it didn't. There are some wax moth larval feces and a few adult bee parts and some propolis specks, but I believe the main component is granulated honey. The bees have a difficult time using the largest granules of honey which has crystallized and drag it out of the hive as debris. These granules dissolve in water, so without doing any other research, I'm going to take a leap and assume that the granules are crystallized honey.

Check your bees to see if they may need some liquid feed to be sure that enough food is available for them since they cannot utilize these granules.

COMMOTION IN FRONT OF THE HIVE

Question - On warm days in early spring I notice much commotion in front of the hive. What is going on?

Answer - *On a bright day after inclement weather, bees which have been confined for some time will often cause an unusual commotion in front of the hive. Bees will be flying in all directions. Watch them for a short time. If concentration of bees remains in the area of the hive, it is probably a play flight. If the colony should be swarming, the concentration of bees in the air moves away from the hive area. If the bees are grabbing each other by the wings and feet and trying to sting each other, it could be robbing. The concentration of bees will be heaviest at the entrance to the hive. If it is robbing, then reduce the entrance to the hive to 2 inches by 1/2 inch high. This will enable the guard bees to protect their hive better. Do not smoke the hive when robbing is in progress as the smoke will drive the guard bees away from the entrance. If you reduce the entrance and bees are still trying to rob, place some loose grass around the entrance to assist the guard bees in protecting their hive.*

Question - I have a question I am hoping you can help me with. Many years ago I kept bees, and now that my life has settled down somewhat, I am very interested in keeping bees again. The problem is, I am a renter in the heart of San Francisco with no space for either hives or to store equipment. There are many rural areas accessible from my home. However, I don't know the best way to go about finding someone willing to let me keep hives on their property. Since bees are such an asset to any farm or large garden, I'm sure that if I knew how to advertise, I would have little problem identifying the right person.

How would you suggest I go about locating land on which to keep my hives? Also, do you know the address of any California or Northern California Beekeepers' Associations?

John Piette
San Francisco, CA

Answer - I'm glad to hear that you have started keeping bees again. I am continually amazed at the superorganism we call a colony of honey bees. Your fascination and appreciation will grow even more than it is now for these special creatures.

Finding a place to keep your bees will be easier than you think. You have several resources at your disposal. Dadant and Sons has two branch office/warehouse locations in California (addresses below). Between our two managers, they can put you in touch with just about any beekeepers in the state. These beekeepers can then give you the inside information on locations available. Contact your state association (addresses below). They will also be able to help. Also remember that any building with a flat roof or gently sloping roof is a candidate for a bee hive. No one will see or become annoyed with them and with the range and foraging ability of honey bees, you will probably also get a modest honey crop even in the city.

Dadant and Sons, Inc.	*Dadant and Sons*
Mr. Pat Eakle, Mgr.	*Mr. John Gomez, Mgr.*
2534 Airways Drive	*2357 Durham Dayton Hwy.*
Fresno, CA 93747	*Durham, CA 95938*
209-292-4666	*916-893-0921*
CA State Beekeepers Association	*San Francisco Beekeepers*
Carol Penner	*Lenore Bravo*
19980 Pine Creek Road	*47 Levant St.*
Red Bluff, CA 96080	*San Francisco, CA 94114*

Question - As an amateur beekeeper, I am interested in the African honey bees. Is there information on how much cold weather they can take? I understand they do not live in the colder areas, such as in Mexico City. Is it known what kind of temperatures would restrict their further spread?

Willard Spankus
Houston, TX

Answer - *You are not the only person to ask this question. Researchers have been very interested in this question because it has a major bearing on how extensively Africanized honey bees could populate northern North America.*

Researchers first put colonies in refrigeration chambers at temperatures of 0°C to -2°C for three months and they survived. Africanized bees have been successfully wintered in natural outdoor settings in Poland, Germany and France. One over wintering study in San Jose, Argentina took place above the 1400 M level in the mountains. Average monthly temperatures reached a minimum of approximately -2°C at the site. To quote from the study, "Colonies of Africanized and European honey bees showed no significant differences in the rate of winter survival."

Based on the above information, it appears that Africanized honey bees will be able to over winter successfully in a great majority of the U.S. I personally am not overly concerned at this because I am confident that the beekeepers in the U.S. are just as capable and intelligent as the beekeepers who have adjusted to the Africanized bee since the late 1950's in all of South and Central America.

P.S. - I was able to use the brand new 1992 Hive and the Honey Bee edition for information in the above question. At 1324 pages, it is the most complete reference guide to all aspects of beekeeping ever made available. This excellent resource is available to the general beekeeping community now. You need to have one available to use at all times. It's great.

**SOMETIMES THE BEES HAVE THEIR
OWN CONSTRUCTION PLANS**

Question - It's extracting time and I need a question answered. It seems like I read somewhere that you could buy Bee-Go that had a different scent. It wasn't like the regular Bee-Go that stinks so bad. Can you tell me anything about it?

Also I want to collect a lot of white sweet clover seed from along the highway. When do you know it's ready to harvest? Should I plant it this fall? If I do will I need to mix an inoculant with it?

Ray Almond
Pocatello, ID

Answer - Your question about how sweet scented Bee-Go works is a good one. There is a company selling the same chemical used in Bee-Go, but they have added a fruit scent to it. The product does smell better for about the first 10 minutes until the fruit scent evaporates away. Then you have the original aroma again. There is nothing that can mask the smell of Bee-Go completely. If you did, then you might not have an effective product like Bee-Go has been.

White sweet clover is an excellent forage plant for honey bees as most know. You should harvest the seed to grow your own stand of sweet clover when the flowers have dried to a dark brown. This would indicate that the seeds have matured and are ready to harvest. Sweet clover should be sowed in the winter or early spring. It should be seeded early enough so that it will have time to establish itself before winter sets in. Remember that sweet clover only lives two years and must regrow from seed, so never cut it when in blossom unless you plan on re-seeding from another source. Sometimes it is difficult to get sweet clover to establish itself, especially on land deficient in lime and you may want to have your soil tested and treated with lime. It wouldn't hurt to inoculate the clover seed with nitrogen-fixing bacteria also. This will assure you of a vigorous nectar- and soil-enhancing crop.

WE HAVE WINNERS!

Dear Mr. Hayes: In the March 1997 *Bee Journal* under your "Brilliant Idea Department", I read the idea about feeding marshmallows to bees. I decided to try it. The only marshmallows I had were the Easter candy peeps and bunnies. The kind that are covered with different colored sugar - yellow, pink, purple, etc. I decided to go with a yellow Bunny and see what happened. Two days after installing it on top of the frames I checked it. The yellow covering was all but gone and the bees were all over the white marshmallows. Seems they loved it. Thanks for the idea.

Troy Mastin
Helper, UT

YOU WIN! The first annual "Brilliant Idea Dept. Grand Prize" selected just for you, a 1997 Dadant Stainless Steel Smoker w/shield.

I especially like the creativity of using different colored Easter Peeps and Bunnys. You're a man after my own heart. There is no reason beekeeping should be solemn or dull. And, adding a little color to hive feeding could just be the ticket. Congratulations again!!!

AND MORE!

I tried using marshmallows to feed one of my colonies and had excellent results. I placed one of the marshmallows on the top bars directly over the cluster. Even on chilly days, the bees continue to eat away at the marshmallows. After observing the bees for several days, I am convinced that this is definitely a better way to feed a hive during the winter. Thanks for tipping us off to such a good idea.

I really enjoy your column and appreciate the effort you put forth to answer our beginner's questions.

Colby McMullen
Williamsville, VA

Thank you for trying out the marshmallow idea. I thought it might work and I'm glad it seems to under real conditions.

You weren't the first to respond back to the challenge to see if marshmallows could be used as an emergency feed, but I'm going to send you a Dadant Smoker anyway. Thank Again.

Question - I just received a spring catalog from a well know seed and nursery company. In it is an advertisement for a bee scent, claiming that use of the scent will increase fruit set and fruit size. Further it "uses the same scent bees do when marking their favorite locations. Makes bees flock to feed on your plants - a great way to boost pollination."

Fred Fulton
Montgomery, AL

Answer - At some time in your life you have probably heard the expression, "you can lead a horse to water, but you can't make him drink". It's kind of the same thing with these bee scents or lures. You can fake them out and "lure" them to an orchard, let's say, but you can't make them collect nectar or pollen. The nectar collection/pollen collecting behavior has other stronger biological reasons and forces behind it than just being some place without a plan (Readers, see Ambrose article in April issue in the <u>Agricultural Research Section.</u>)

The question of marking flowers is a good one. Many insects that feed on nectar and/or pollen leave behind a scent not to mark the location for other insects that are of the same species to visit, but to tell them not to visit this flower. It makes sense for survival of the species to tell this sister insect not to waste her time at this flower because all the good stuff has already been eaten. Likewise, some flowers when pollination or fertilization has been accomplished, send out a scent to say insect visits are not needed any longer.

I have not read either whether honey bees mark flowers with a <u>do</u> or <u>do not</u> pheromone. If I were to guess though, I think at some level they do. It is the most efficient way of collecting the most nectar and pollen which ties into survival of the species.

Question - I produce both honey and pollen in western Colorado. I retail the honey through local stores, and I wholesale the pollen. Some local people who have found out I produce pollen have become regular purchasers of pollen. The question I have is this. Is there a standard and legally sound disclaimer I could print up and include when I sell pollen to individuals that would protect me legally from adverse physical reactions they might experience and from any implied claim to medical benefits?

James Tipton
Glade Park, Co

Answer - As much as I personally dislike the parasitic nature of attorneys, sometimes they are the lesser of greater problems. We in our current society love to sue our neighbor. Every day one can read about someone who had something frivolous happen to them and will sue whoever they think is responsible. By the same token, there are legitimate reasons for a person who has been harmed by machinery, medicine, etc. that was improperly made or administered to try to get redress.

My suggestion is to spend a little money now to consult with an attorney familiar with the food/pharmaceutical industry to save maybe a lot of money later. Check with your insurance agent to get his/her input and input from the company. Is your current liability insurance enough? Will they cancel your insurance if you start selling pollen without what they feel is proper labeling and disclaimer? How is your business registered? Is it a sole proprietorship, corporation, partnership, etc.? How it is set up can also determine when you have a suit filed against you if you are personally liable of just your business. Remember that in our country of 250+ million people, that there is at least one customer of your products who may have a slight to severe allergic reaction to your pollen product through no fault of your own. If they sue, it is only prudent and sensible that you are ready. Don't ever assume that because you are a decent upstanding guy that someone would not want to finance his or her retirement off you.

My advice is to contact the people who are legal and insurance professionals to tell you how you should protect yourself, your family and your business.

HOUSE APIARIES

Question - In your August 1990 issue, you had an article from 100 years ago about a house apiary. I found this article quite interesting and I would like to know more about this concept. As a hobbyist beekeeper, I have a few questions that you might be able to answer:

1) Is this concept in use today?
2) Are there plans available for the construction of one?
3) Could this concept be adapted to using it on a trailer?

This would allow the easy transportation of a number of hives (permanently located in the house apiary) from a wintering area, to a field of clover or alfalfa in the apiary, then, to a field of sunflowers in the summer, then back to the wintering area in the fall, with little or no disturbance to the hives. NOTE: This concept could also allow for storage of equipment in the house apiary.

Thank you for any assistance you can give me concerning this idea.

Rod Kensrud
Fargo, ND

Answer - The house apiary is in wide use throughout many parts of Europe and the Mediterranean. They range from simple open front sheds to elaborate multi-story structures with extracting rooms and workshops inside. There are no construction plans that I know that are readily available.* The main points to follow are a bee-tight structure so that the only way honey bees can enter their colony is through the entrance given. Generally, there are windows and electric lights; some have running water and all the necessities of home. The bee house is simply a building in which colonies are brought in and pushed up next to the wall provided with a hole at the same height as the entrance on the colony and thus the hives are protected from the elements and vandals. As you noted, some of the bee houses are on wheels and moved when needed.

The design is left up to the builder. Have fun with this concept that works in many places of the world.

*John Spiller wrote a book entitled <u>The House Apiary</u> in 1952. Printed by W. I. Cornwall & Company of Exeter, United Kingdom, it has long been out of print. We have two copies, however, in the Dadant Apicultural Library. It has plans for a house apiary building, but not a mobile apiary. However, European beekeeping journals regularly publish pictures of mobile house apiaries. We suggest you write to some of these journals. A few of their addresses are listed under Periodicals in the Marketplace section of the American Bee Journal.

Question - While visiting the U.S. recently, I came to know some beekeepers. They used your catalog exclusively when ordering any supplies and equipment. I thought this very interesting that such a big country would only have one beekeeping supply company. Is this true?

Sebastian Van Nes
Emmeloord, Netherlands

Answer - The U.S. is a big country and even though Dadant & Sons have grown to be the most well know beekeeping supply company, we are not the only one serving the industry. There are many regional beekeeping supply companies that do a very good job of providing supplies and equipment to beekeepers in their area. Dadant & Sons with its sophisticated manufacturing capabilities, 10 company branch warehouse locations and over 600 independent Dadant and Sons distributors, does have the ability to reach most beekeepers in North America.

The story of the humble beginnings of the French immigrant, Charles Dadant, coming to America in 1863 and building the pre-eminent beekeeping supply company that it is, is one of those unique American success stories. And, it's heartening to us "Baby Boomers" that he didn't even start his beekeeping business until his late forties.

Question - I have two hives placed less than 100 feet apart. One is a feral colony that I captured 3 years ago and the other is a 2-pound package with Midnite queen that I started this spring. Could you tell me why the new colony produced much lighter colored honey than the old one?

Leo Matt
Taberg, NY

Answer - I think one answer as to why your two colonies produced different types of honey is that they were simply foraging on different flowers. Remember that both colonies will be sending out scouts to find the best nectar source. These scouts go and sample nectar from as far away as a mile or so. The scouts could have easily gone in different directions, found different nectar sources and recruited foragers to go to these different places. Unless the nectar source is very large and productive, honey bees do not like to compete with other insects or bees. One colony's scouts could have found the same field of flowers as the second colony's, but because these honey bees were already there, the nectar samples these scouts picked up would have been very small and unimpressive to the colony's recruits. They decided to look for something better.

This type of thing happens more often than one would think. It's kind of neat that you are in a location where you can get two kinds of honey from two colonies.

The other answer to the difference in honey color is that the older colony simply had darker combs (possibly used for brood rearing). These darker combs produce a darker honey, despite the fact that the nectar may have been just as light as that which was gathered by the newer colony. This is one reason beekeepers who strive for lighter color honeys are urged to use queen excluders and replace darker, old super combs as necessary.

Question - Has any research been done in the U.S. about the hive tone or sound levels?

Rice Mace
Valley City, OH

Answer - There has been much solid research done on the sounds coming from hives for 40 or 50 years. Equipment, process and procedures have been developed to identify queen sounds, wax moth sounds, workers preparing to swarm sounds, sound differences between Africanized and European honey bees and any other sound you can think of. The problem with the commercialization of this equipment has been that the beekeeping industry has never sensed any of these concerns as a problem that many beekeepers would spend money on to identify. Without a real problem to be fixed, the research never gets out to the public.

Although, for example, it might be nice to know when your hive is queenless or preparing to swarm simply by listening to and interpreting the sounds emitted, there are methods available now that are just as reliable and a lot less expensive. The commercial beekeeper or long-time hobbyist will be able to tell when a hive has a problem simply by looking at the flight activity at the hive entrance or making a quick check of the brood chamber. Perhaps these sound instruments will be used by beekeepers some day, but right now they are still very expensive and not entirely reliable.

A NICE FRAME OF BROOD

Question - I am hoping you may be able to settle a dispute that my girlfriend and I have been having concerning the wintering of bees outdoors. I have been trying to impress upon her the importance of keeping the snow shoveled away from bee hives so that the bees will be able to get outside during the winter months.

She maintains that this is preposterous and unnecessary, as the bees are hibernating. The piece by Charles Dadant on page 569 of your Sept. 1994 issue *Wintering Bees Outdoors*, seems to agree with my position relative to the importance of providing egress for bees during the winter.

If you could put this discussion to rest, once and for all, I would be very appreciative.

James Madden
Iowa City, IA

Answer - *It's nice to be able to say that both of you are partially correct in your assertions regarding snow around beehives.*

Snow is a good insulator from cold temperatures and cold winds during the winter time. Therefore, many northern beekeepers who winter their bees outside prefer a winter with enough snow cover to insulate the hives.

However, bees do occasionally need to take cleansing flight to rid themselves of accumulated feces. For that reason, it is advisable to at least remove the snow from in front of the entrances of the hives. This will provide an exit for the bees and in some heavy snow or ice situations, the removal of the snow in front of the hive will also prevent bee suffocation. Nevertheless, it is perfectly fine and probably advisable to allow the snow to remain on the other three sides of the hive to act as an insulator against the cold weather and wind.

Bees need to take cleansing flights at least every month and preferably more often than that. Unfortunately, this isn't always possible in the northern climates where extended periods of below zero weather prevent regular cleansing flights. In some cases, this results in severe bee dysentery.

The main thing to be sure of is that your colonies have plenty of honey to over winter and that you check them from time to time to correct any problems such as ice or snow coverage of the entrance or lack of stores above the winter cluster.

By the way, honey bees do not "hibernate". They form a tight cluster during extreme cold to conserve heat and food reserves. Even though this cluster of bees does not venture outside except during warm weather, it could never be called "inactive" or "hibernating". In fact, limited brood rearing often continues throughout the winter.

Question - In the January 1997 issue of *ABJ* heating hives by electricity was mentioned. (50 years ago)

As I recall, a company was advertising electric heaters for bee hives back in the 70s.

Obviously, for some reason, they are no longer available for sale. Were they effective and/or economical? Would advances in technology since the 1970s make them practical today?

Rufus Payne
Appalachia, VA

Answer - *Hive heaters or other ways to retain heat in a colony have made appearances off and on for years. Yes, heaters will add heat to a hive, but in temperature climate hives this creates its own problems and no it's never economical to use any form of energy to heat a beehive based on the retail price of honey.*

First, honey bees. Apis mellifera when free of diseases and parasites and having adequate stores of honey to eat, can withstand the best that North American winters can dish out without any additional help from the beekeeper. If heat is added to a colony of bees they, of course, become more active, eat more, have a need to defecate more and start raising brood which adds to the pollen/honey food reserve problem.

Under normal field conditions heating a colony of honey bees is never efficient or economical.

Question - At our last local beekeepers meeting, we had a speaker talking to us about smokers and their proper use. One of the parts of the talk was about what you could burn in the smoker for a cool white smoke. Some of the things were, standard like corn cobs, rotten wood, pine needles, dry leaves and such. Some of the stuff talked about was new to me, like adding high nitrogen fertilizer to produce nitrous oxide or laughing gas, old oil filter cartridges, dried mushrooms or puffballs and some other things I won't mention here.

My question is what is the best thing to burn in my smoker and are some of these things safe to use?

Brian Rhodes
Beaver City, NE

Answer - I've always been uncomfortable with the use of smokers because even when used properly to calm a hive, I've never liked the smoke, no matter how good it smells as when using pine needles. I'm not a cigarette, cigar or pipe smoker because of all the reasons we all know not to smoke, so I've never really enjoyed using a smoker even when I have to. I try to use a smoker as little as possible, but realize that its use is invaluable at times and is a vital part of beekeeping at this time and probably in the future unless some alternative is found.

First, let me answer why does smoke work to calm honey bees? Honey bees communicate primarily by the use of pheromones. These are chemicals released by the queen or workers which, when smelled or tasted, tell the bees about what's going on in and around the colony. When you puff smoke in the colony, it overwhelms the pheromone odor. The bees now cannot communicate to each other; they are individually isolated. Without any clue to what's going on and all organization temporarily breaking down, the bees become less defensive, making it easier for the beekeeper to "work" the colony. The smoke also causes the bees to ingest large amounts of honey, as if absconding, and these honey-filled bees are not as likely to sting.

Unless you wear a gas mask, you are going to inhale some of the smoke. If you have one colony, you will obviously inhale less smoke than if you have 100 hives and are out working them for hours. Not even addressing what the smoke does to bees, in addition to "calming" them, do you want to put high nitrogen fertilizer in your smoker to burn and inhale nitrous oxide also? In addition to nitrous oxide, the poisonous gas hydrogen cyanide is also produced from burning nitrogen fertilizer. You surely don't want to breathe that! Or how about the old oil filter cartridges burning and inhaling goodness knows what in that smoke! The one you mentioned about mushrooms or puffballs really made me shake my head because some species of these contain chemicals in them that result in a hallucinogenic effect on people.

The narcotic effect on bees may be due to hydrogen sulfide, another gas people want to avoid inhaling! Sounds to me that your speaker has been standing too close to his smoker while burning mushrooms for too many seasons. I think his comments were irresponsible.

A smoker can be a valuable beekeeping tool when used properly. Burn pine needles, dry leaves off of known trees, straw, fine wood chips, etc. Remember to use your smoker sparingly because if too much smoke isn't good for you, it probably isn't good for the bees. And use common sense in what you burn in the smoker.

Question - I'm just a hobby beekeeper, but like most beekeepers, I have been lucky enough to produce more honey than my family can eat. I'd like to sell some, but am concerned that I need some kind of license or health department inspection. Would I need special labels? How can I do this legally?

Dianne Behnke
Kahoka, MO

Answer - First, you are not just a hobby beekeeper. The only difference between you and the commercial guy is the number of colonies you have. Your goals are the same. Your basic equipment is the same and many of your problems are the same. Don't apologize because you have fewer hives than someone else.

In the past anybody who wanted to sell a few hundred pounds of honey out of their home could do so without hassles. In many parts of the country you can probably still do so. Let me add a disclaimer and a personal fear I have about this.

Let's assume that you handle, process and pack your honey in a very professional and sanitary manner. Everything you do is perfect. Say you sell some of your honey to someone who is very pleasant and appreciative of your product. They, in turn, consume your honey and have a violent allergic reaction to it that requires medical attention. Their contention is that the product you sold them was defective in some way and that you did not warn them that there was a possible danger in eating your product. And, you did not tell them that you were not licensed to sell food products. With some basic investigation by their attorney, he/she further contends that your product has traces of the antibiotic Terramycin which is commonly used to treat bee diseases. This product shouldn't be in the honey, and, they say that the miticide, fluvalinate, also is present as a residue which shows you misused these products. This resulted in higher residues and caused the health emergency in his client. The customer is going to sue your socks off and probably win with a jury or a judge who doesn't know a honey bee from a cockroach.

I hate to be the bearer of "gloom and doom" especially in an activity in which I participate. But, the reality is that our world and the people in it are less forgiving and more self-centered than in the past. That does not mean that we cannot or should not practice our craft or strive to produce the finest product around. It does mean though that we should follow the letter of the law and not assume that because we are involved in a quaint pursuit that in this day and age we can proceed with a wink and a nod.

Check local and state laws and ordinances. Talk to local packers. Check with your insurance company. Then make an informed decision of what you should do. Or, you can take the easy way out and eat all the honey you produce, like my family does.

In addition to the above considerations, be sure to order your labels from a company that knows what items must be on them and how they should be printed. The new nutrition labels must also be used if you make nutritional claims on your label or have a sales volume over $500,000 per year and of this amount, food sales accounted for over $50,000. Consult the National Honey Board for details: The National Honey Board, 421 21st Ave. #203, Longmont, CO 80501-1421, Phone (303) 776-2337 and Fax (303) 776-1177.

Question - I have been a hobby beekeeper for several years with interest which goes beyond the honey bee. For example, the March issue of the American Bee Journal had an article about a Polish beehive made of Styrofoam. Do you know if this hive or others of alternative material or design are available in the U.S.?

The same issue has an article about stingless bees in Australia. Can you tell me if this or other stingless social bees are in the U.S.?

Victor Peterson

Answer - 1) Styrofoam or similar material hives have not caught the imagination of beekeepers in the U.S. as yet. Cost is still a question because a purely Styrofoam hive cannot be used. The Styrofoam has to be sandwiched between two harder layers of material or the bees will chew right through it. No U.S. company is making or widely distributing anything like this now.

2) There are many species of stingless bees in Central and South America. The Native people of these areas had a highly developed system of "beekeeping", using the stingless bees at one time. They do not produce honey in the quality or quantity of the European honey bees that we are accustomed to.

They may be stingless, but they are not defenseless. Some of these stingless bees spit formic acid which can be rather unpleasant when smeared in eyes or breathed. Also, some stingless bees bite!

They cannot live in our temperature climate since they are tropical bees.

Question - I was considering getting some guinea fowl for my place. I was told that this is not a good idea because the Guinea would sit by my beehives and devour the bees by the hundreds. Have you heard anything on this and can they co-exist?

Tim Armstrong
Ashtabula, OH

Answer - Guinea fowl are omnivorous, so they could feed on honey bees. I have never heard of severe losses of honey bees from anyone having Guinea Fowl chickens, turkey, etc., that were in close proximity to bee hives. Now this assumes that the birds are getting a full adequate diet and not being forced to feed on honey bees as a secondary food source. I think the bee's stinger and venom would deter the Guinea Fowl from consuming too many bees. In fact, years ago there was a study done feeding turkeys honey bees that were depopulated from northern commercial hives going into winter and the turkeys lost weight because they reduced their feeding. They didn't like honey bees as a steady diet!

AFTER-SWARMS

Question - I am a beginner with only six colonies, two with Starline queens and four with dark Italian queens. At the beginning of last spring, I had two colonies, one with a Starline, the other dark Italian. The Starlines swarmed once, the swarm weighing about 9 pounds. The dark Italians swarmed three times, once in late April, again in early June and again July 15th. The swarms weighed 8, 10, and 15 pounds and yet the hive is still full of bees and emerging brood. Should I attempt to stop these after-swarms?

Answer - You should try to stop not only the after-swarms, but also the prime swarm. It is best to keep bees from swarming because the queen, a large portion of the bees, and part of the honey goes with the swarm. Also, there is a period when the colony is left without a laying queen. Swarming may be controlled in part by early and adequate supers, good queens and combs.

Question - Mr. Hayes, could you please tell me the status of the BANATS honey bees? Has it been reclassified into another strain, or is it now extinct?

R. S. Amick
Double Springs, AL

Answer - I found additional references on the Banat Bee between 1908 and 1915 in primarily ABC XYZ published by Root. Their inclusion as a race of bees was either not known or unappreciated by other authors of the day. The word banat or banate is of Serbro-Croation origin which makes sense since it appears that this bee resembles Carniolans and Caucasians in appearance and action. The description of these bees is exactly like that of Caucasians.

I checked with some contacts in Europe and they were not familiar with the term Banat as it relates to honey bees. Since it looked like a Caucasian and acted like a Caucasian, I'm not quite sure why for a short time in the bee literature it was classified separately. Someone must have thought better for classification of it as race seems to have disappeared as a reference item.

Question - Can you get me the plans to construct a triangular bee escape board, the one that works on a maze principle? It seems that I have seen them described in the *ABJ* before, but I am unable to locate one now. I am told that this type of escape board works much better than the Porter bee escape.

Also, I would like for you to describe the method of queen rearing that one uses by allowing the queen to lay in a frame and then destroy cells horizontal and vertical, leaving every third cell and then laying the frame flat over the top bars of a queenless colony with enough space for cells to be built.

James Neagle, Jr.
Richmond, VA

Answer - Two good questions. Thanks. I'm stumped on the first one, though. The triangular bee escape board was, if I remember correctly, developed in Canada. I have seen one and I think you are right that there was an article in either the ABJ or Gleanings years ago. Anyway, I couldn't find any picture or description in our Dadant library. If some reader could help, that would be great. If not, contact Dave Heilman at OSU/ARI, 1650 Madison Ave., Wooster, OH 44691. He has one in the Archives there of which he may be able to take some measurements.

The second question is easier because I wrote an article years ago on the Hopkins Method of Queen Rearing. I still think it is a neat way for the hobbyist to do it. I'll send a copy of it to you because it's too long to reprint it here. If anyone else would like a copy, drop me a note and I'll send one out.

Question - I'm looking at raising bees for honey and profits as a sideline income. I have a garage full of Philippine Mahogany and am wondering if this would make good hives, supers, etc.

Patrick Rheaume
Mecosta, MI

Answer - With the price of lumber being so expensive nowadays, especially the more exotic woods such as mahogany, I certainly wouldn't use it for making hives which are usually made of cheaper wood such as loblolly pine or white pine.

Some beekeepers do make hives and hive parts and sell them, but the cost is rather prohibitive unless the maker does not count his labor.

Larger manufacturers of hive equipment have high speed modern equipment to accomplish this task at a much cheaper per item rate. I think once you put a pencil to it that it will be much cheaper to purchase most of your equipment from one of the mass producers of woodenware such as Dadant and Sons.

Question - My father and I have approximately 75 hives. We are getting tired of trying to store, uncap, and extract our honey in a combination of our garage and kitchen. We would like to build a honey house, but need some idea of the optimum way to set up equipment in it first. Do you have any suggestions that we could secure for our guidance as we set up our "honey house?"

LaRue Stevens
Jackson, MS

Answer **-** *As you have discovered, it gets pretty aggravating and frustrating not having a well laid out honey house when you get to a certain number of producing colonies.*

You are absolutely right in researching the best honey house equipment layouts and designs before construction. You will have to live with your decisions for a long time, so you might as well take time now.

I would suggest that you read chapter 15 in the latest edition of <u>The Hive and the Honey Bee</u>. *The whole chapter deals with honey and wax processing. There is a wealth of suggestions concerning equipment, its use and placement. There are several pages of floor plans for you to review.*

Also, talk to larger beekeepers in your area who already have honey houses. They can surely give you information from the "horse's mouth" about what works and what doesn't and how they would do it differently next time.

I've included just one of the many possible designs and layouts that one could use. There are many, many others that are efficient also. You may adjust the size of the extractor and tanks, depending on the size of your operation.

Question - I have read a number of books and watched films on beekeeping, but would like to learn more. Is there a school where I might get hands-on experience? Thanks.

Joseph Farley
Broken Bow, OK

Answer **-** *I am glad to hear of your interest in beekeeping and of your desire to know even more. Hands-on experience is vital for a thorough and confident grasp of any subject, beekeeping perhaps more so than others.*

My suggestion is to first contact the Oklahoma State Beekeepers' Association. If anyone would be able to tell you about short courses or college level entomology courses available in the State of Oklahoma they should. Contact Dorothy Smith, P.O. Box 34, Guthrie, OK 74044; Phone 405-282-4002.

For actual beekeeping schools with degree programs in practical beekeeping, contact: Dr. James Tew, ATI - Ohio State University, 1328 Dover Road, Wooster, Ohio 44691; Phone (216) 263-3684 or Fairview College, Beekeeper Technician Program, Box 3000, Fairview, Alberta TOH ILO CANADA (403) 835-6000.

Question - 1) In western Washington, our winters are wet and cool. This climate leads to a very damp environment and often a damp hive. I understand that I should dry my hives, but when I open them for ventilation, I face the possibility of getting the hives too cold. How damp is too damp? How cold is too cold?

2) The caps and trimmings I melt down for candles. But the dark comb doesn't melt well and usually leaves a mess. Can this comb be used for candles and if so, how?

Don Lorimor
Port Orchard, WA

Answer - *How damp is too damp and how cold is too cold? I've never heard of any numbers on this nor have I been able to find and research numbers in my quick scan of the literature.*

Bees can withstand and control low humidity cold better than high humidity cold. For instance in the Midwest it gets very cold on occasion. Recently it was 20° below zero. Humidity drops rapidly below freezing because the water (humidity) in the air freezes. It is only above freezing when humidity and cold can cause water to actually collect in the hive. Giving some upper ventilation is about the only thing you can do. The rest depends on your strain of honey bees being able to modify their environment to deal with it and the stress diseases that are opportunistic in these conditions.

Beeswax is darkened by larval skins for the most part and pollen and other debris that over time is incorporated in otherwise light beeswax. In fact, the bees remove wax over time from dark comb, using it elsewhere and leaving the dark material with less wax than extraneous material. The wax remaining in this old comb can be removed to some degree by putting it in a clean porous cloth bag and submerging it under boiling water for several hours. The wax separates from the other material, works its way through the cloth bag and floats to the top of the boiling water. When the water is left to cool, a large layer of cleaner beeswax will collect on the surface of the water. It won't be as clean as new wax capping, but it will be more workable than you are experiencing now.

Question - Some of the regular writers in the *ABJ* have their picture with each column. How come you don't?

G. Pyle
Mayberry, NC

Answer - *The writers of the columns who included their pictures have done so because they are real people. In order to save money, Mr. Joe Graham, the editor of the Bee Journal, has decided to use virtual writers. I and others are really computer generated. We all have separate personalities and writing styles, but are not flesh and blood. We only exist as bits of electrons in a computer. The savings realized by doing it this way keeps the subscription costs of the Bee Journal down so more people can learn about beekeeping. F-4 and off.*

Question - I'm on a low fat diet and was wondering if the honey, pollen and beeswax I eat have any fat in them.

Iris Cooley
Metairie, LA

Answer - *There are some excellent sections in the new edition of The Hive and the Honey Bee that give you complete nutritional breakdowns for these products.*

In a nutshell, this is what they contain. Honey is predominantly various sugars with some vitamins, minerals, enzymes, etc. but no fat. Pollen has vitamins, minerals, proteins and fat (lipids). The fat content of pollen is about that of the same amount of chicken. Beeswax is made up of lipids and hydrocarbons, but is generally considered indigestible. In other words, the beeswax passes through your body and the fats are not added to your system.

Question - I am very interested in beekeeping, but know very little about bees. If there is any type of book that would tell a person everything there is to know about beekeeping and more, I would be very interested. I currently subscribe to the *American Bee Journal*, but need something that would help me get started.

Matt Litwiller
Goshen, IN

Answer - The first and last book that you will need is "The Hive and the Honey Bee". This has been extensively revised recently and has virtually all current information available to the reader at his/her finger tips. There are 27 chapters, 33 authors and 1324 pages. At $36.00 this is the bargain of the year.

I strongly encourage you to purchase this book. It will give you either quick reference or in depth information for years to come. It's great.

Take care and good luck with your new interest in beekeeping.

Question - A friend recently sent me a book, *Beekeeping in South Africa* which is of course about African bees. On page 13, it discusses the different brood cell sizes as compared to European cells. I quote, "These bees because they use smaller cells do not readily accept the larger European cell type Foundation."

"This is especially noticeable when hived on frames with European size cells. In fact they tend to abscond..."

As secretary of the Houston Beekeepers Association, I am very interested in the problem if it exists with the bees that will be swarming into Houston in a few months.

Stuart Kuik
Houston, TX

Answer - The researchers at the USDA who would know about this indicated that they had never heard or seen research documentation that showed that African bees in Africa would abscond in response to European cell size foundation. They have never seen or heard of great problems with Africanized bees from South and Central America either.

I was told that the Africanized honey bees moving up from Mexico took longer to accept European cell sized foundation and/or comb, but absconding did not take place. The Africanized honey bees simply reworked and made the modifications necessary for what their needs were. The foundation or comb was readily accepted to the extent that the bees stayed and made whatever changes to the structure they wanted.

Question - I am a beekeeper who buys packages every spring and kills them off in the fall. I have been using Cyano Gas to kill them, but now I cannot buy it anymore.

Do you have any suggestions how to kill them some other way or do you have anything to replace Cyano gas?

Clifford Eggen
Thief River Falls, MN 56701

Answer - As you found out Cyano Gas is no longer available. It hasn't been approved for use in U.S. beekeeping for many years. The cyanide produced by this product was very effective in killing bees and beekeepers!

There is no government-approved chemical way to kill off bees that doesn't also make unusable the beeswax comb and honey left. Some commercial beekeepers have experimented with car exhaust fumes as well as bee vacuums that grind up the bees into fertilizer. However, both of these methods have their drawbacks.

With the cost of package bees going up and up, most beekeepers who used to kill off their bees every fall are now finding it more cost effective to overwinter their bees. I would put a pencil to this first and see if it would work for you. If not, just extract all the honey and let them starve to death over winter. Sounds cruel and it probably is, but is also cheap and legal.

Question - 1) Solar Wax melter question: If the frames could be put in the melter upside down (vertical), would this speed up the wax flow?

2) The solar Wax Melter will have to be deeper to handle brood frames standing. Is this going to hinder its operation? –Any plans?

3) I would like to build a trailer that would carry about 10 hives, the hives would be permanently left on the trailer. What are the disadvantages of using a trailer? (the hives will be extremely close)

4) Most of the moves will be 4-10 miles. Have you seen any well thought out trailers?

5) Would you make the trailer long for one row of hives 5' x 20' or 8' x 16' with two rows of hives? (I would face all the hives out.)

6) I made cedar bottom boards for the hives this year trying to ward off moths. The bees seemed to cover the cedar smell by the end of July. I intend to put fresh cedar saw dust around the entrances next summer. Have you had any experience with this?

7) What type of paper towel would you prefer to use for applying vegetable oil as a mite treatment, should it be extra tough or very thin?

8) In my last letter I asked about using nuc boxes to start new swarm. After reading several articles I now think I should be using 4-frame starter boxes. If you were building starter boxes using brood frames, is there anything you would change from the standard box?

9) Do you have an E-mail address?

Myron Denny
Stillwater, OK

Answer - 1) There is an old movie that parodies an efficiency expert who is always timing everything he does including buttoning his shirt from the bottom up or the top down. He determines buttoning from the bottom up is faster and more efficient. Like this story, any difference in wax flow would be so small as to be inconsequential.

2) The only thing that will hinder your solar wax melter operation is the heat you can capture and maintain, not size or depth.

3, 4 & 5) I spoke late last year at the Alabama State Beekeepers and at one of the Field Day sessions, Dr. Jim Tes, OSU/ATI had a neat small bee trailer. Contact him at Ph. 330-263-3684 or Fax 330-262-2720.

6) As you notice, the bees do not like aromatic products in their hive. They cover these materials with propolis when possible. Even if every part of wood your hive was made of was Cedar, the aroma would dissipate naturally or be covered by the bees quickly. The wax moths would not be repelled. I don't think Cedar sawdust outside will reach the aroma concentration needed to keep wax moths out. But, give it a try and report back to us how it went. The best way to keep wax moths out of a living colony is to have a <u>healthy</u> and <u>populous</u> colony to protect itself.

7) You want the bees to come in contact with the oil as they remove the paper towel. Just a regular paper towel, as used in your kitchen, is fine. Some beekeepers buy shop towels in rolls.

8) My problem with either 4 or 5 frame nuc boxes was always a neat quick way to open and close the entrance and keep the top on for ventilation when transporting. Any other changes that help you, but do not disrupt the bees gravity.

9) class@dadant.com

Question - What can I use to kill the bees in the fall? I want the honey to still be O.K. to eat.

Charles Damron
Wyandotte, MI

Answer - To kill bees in the fall and not contaminate the honey is possible but not easy nor completely safe.

Common sense tells us that most any chemical insecticide would leave harmful residues in the wax and honey, making it unfit for human or bee use. In the past here in this country and others, cyanide was used to kill bees. Because of its volatile nature, it would leave no residues. Unfortunately, the cyanide gas kills beekeepers just as well as bees, so access to this poison is now limited. Other less noxious and less quickly effective gases such as carbon dioxide and nitrogen gas can be used to in effect suffocate honey bees. Another technique is to simply extract every ounce of honey and let the bees starve to death. We have even heard of some beekeepers who use strong vacuum cleaners which also kill the bees and deposit them in a canister. The contents are dumped in the garden as fertilizer.

I don't like any of the above means. I am a hobby beekeeper. I truly enjoy my relationship to my bees. Some may think this strange to admit. I would admit my awe and respect for this group of insects. I suppose if I had hundreds or thousands of colonies of honey bees and it was economically more feasible to kill off the colonies in the fall and repopulate with packages in the spring, then my attitude might change or it might not. I'm glad that I don't have to be as mechanical with my beekeeping as some must or choose to be. I guess I'm a softy, but I like them.

If you feel you must do something with your bees in the fall, sell them to someone to boost their colonies going into winter or give them away for the same reason. They can be preserved and you still have the honey. Enjoy.

SKUNKS EATING BEES

Question - I'm having real troubles with skunks eating up my bees. They scratch on the hives, the bees come out to defend themselves and they are eaten up. What can I do?

Maurice Ludington
Valparaiso, IN

Answer - Skunks can be a real problem in many parts of the country. They can weaken a colony in a short period of time.

Skunks are opportunists, like many creatures, and once they find out that by scratching at the hive entrance at night the bees march out to be gobbled up, it's hard to get rid of them. It's a real easy meal, a veritable buffet for the taking.

I've checked the literature and there have been many attempts to trap, kill or discourage skunks from feeding on honey bees. Poison, various stinking chemicals, traps, cages, drowning, fencing, etc. have all been tried. Live trapping and moving the skunks many miles away seems to work as does outright shooting of the skunks.

Some beekeepers claim that if nails are pounded through a board and then the board is inverted in front of the entrance, the skunks will be discouraged from approaching the hive entrance. The sharpness of the nails sticking up through the board does not provide a welcome mat for the skunks. Be careful that you don't also step on the "skunk board".

I had skunk problems myself several years ago and used the following method of discouraging the skunks so they would go elsewhere. I sealed the bottom entrance and created an entrance above the supers. When the skunks scratched on the hive, the bees would come out at the upper entrance which was too high for the skunk to reach and it couldn't feed. Skunks are intelligent and resourceful so if this doesn't work, you may have to try harsher measures.

MASTER BEEKEEPER

Question - Often I have heard reference to beekeeper as a Master Beekeeper! How does a beekeeper attain the title of Master Beekeeper?

Choron Grimsley
Center, TX

Answer - *The "Master Beekeeper" program, supported by the Eastern Apicultural Society, awards the Master Beekeeper title to beekeepers passing a series of rigorous written and laboratory tests and an in-depth practical field exam.*

Anyone wishing to obtain information on next year's testing program should contact Dr. John Ambrose, North Carolina State University, Box 7626, Raleigh, N.C. 27695-7626. Phone (919) 515-3183.

Question - There seems to be no life in my weakest hive since we had a temperature of 25 degrees below zero. When can I open it up and see? If the bees are dead, can I leave the combs as they are until I get a 3-lb. package April 1st? Could I use the honey in the combs if there is any left? Do I need to seal the hive to prevent wax moth damage?

Steve Martin
Nauvoo, IL

Answer - *It would be desirable for you to wait until the temperature gets to 40 degrees above zero before opening the colony for inspection. The warmer, the better, of course. By opening the colony in very cold weather, the heat within the colony escapes, and you endanger the survival of the colony.*

If in fact the bees are dead, the combs should be cleaned of dead bees to have them ready for the package bees. The honey that is left could be used by the package bees.

There will not be much danger of wax moth in the hives now as the wax moth doesn't really operate until the temperature warms up considerably. The invasion by mice would be of more danger since they would quickly destroy the combs. Another problem may be robber bees trying to rob out what honey is left on warm days.

Question - A few of my supers have some drone cells and bees don't put honey in those cells. Is the only solution to replace with new foundation?

I enjoy your articles in *ABJ* and benefit from them. Your comments would be appreciated.

James Booth
Mobile, AL

Answer - *The worker bees will choose worker size cells over others to store nectar/honey in first. If they use these up as storage spaces, then they will use drone size cells to store honey in. In fact some beekeepers purchase a special foundation from Dadant and Sons that has drone cells embossed on it so it will be drawn out into drone size cells only. These are then used exclusively as the honey supers. The theory behind this is that since the cells are bigger and require somewhat less wax to build that more honey can be stored and that because of the larger size cells it's easier to extract. I haven't tried this myself, but I can't argue with beekeepers who have and experienced success.*

This doesn't really answer your question with a yes or no, right or wrong answer. It's kind of like all beekeeping where there are larger areas of additional questions. I'll leave the replacement with new foundation up to you.

Question - Although overwintering of bee hives in cellars is not as prevalent as it used to be, how is this possible when cleansing flights are impractical in the confined cellar for 3 or 4 months? Also, how is the condensation problem resolved in this situation?

I ordered a commercial queen by mail in October. She was slim and did not appear to be mated. About ten days after installation, she began laying. However, I noted there were two eggs in some cells! Isn't this unusual from a new queen? Do both eggs ever develop into the pupa stage within the same cell?

There have been times when handling a queen that she would go limp on me, as if feigning dead, and would recover about 20 seconds later. Is this a self-preservation measure? Have you heard of this happening?

Fred Fulton
Montgomery, AL

Answer - *Overwintering in cellars is not as prevalent as it once was. You are absolutely right. However, environmentally controlled modern overwintering buildings are popular in Canada. Temperature, air exchange and humidity could not be controlled in cellars as they now can be. At the right temperature, and with a diet of light honey or fructose (no extraneous matter to clog intestines) and freedom from Nosema, honey bees can go without cleansing flights for three or four months, sometimes longer. I know, I used to live in western Michigan where my colonies would be buried in snow for many months at a time. There are several pages in the new edition of The Hive and the Honey Bee which describes indoor wintering in great depth. Indoor wintering is quite interesting and in northern climates can be more efficient than outdoor overwintering.*

With a newly mated queen who has been trapped after mating, put in a little box and sent through the mail, she has no reason, desire or location which tells her to start making eggs which give her that plump appearance. Queens start slimming down in fall and winter also as conditions for laying eggs deteriorate. This is a perfectly normal condition for a newly mated queen. Newly mated queens also have somewhat of a physiological learning curve to go through. That's why sometimes on new queens you may see more than one egg in a cell. She is learning, if you will, about laying eggs and how to put only one egg in a cell, whether fertilized or not. When workers who are feeding and tending these cells realize this, everything is removed from the cell and it is cleaned in preparation for another try.

There is also the possibility that your new queen was a "drone layer" or improperly mated queen. She also may have been injured or diseased. Keep checking on your queen. If her laying does not improve by this spring, requeen or unite this colony with a "queenright" colony.

Some queens or lines of queens from certain mothers are genetically programmed to "faint". This probably is a weakness in breeding (inbreeding) which shows itself in this way. The queens will revive and lay normal patterns. It is an interesting phenomenon though and you got to see it close up!

Question - I wish to ask the following questions:

1) Why do different grades of honey exist on the market? It's all honey produced by the same insect *Apis mellifera.* I know well that the color and the moisture depend on the type of flower, where the nectar is collected and the timing of the rains. But, my question is what other factors contribute to the existence of different honey grades?

1A) What types or grades of honey are normally recognized by the U.S. government?

2) In order of production, what race of honey bee is most efficient in production of honey, both quality and quantity?

2A) I know there are different natural colors of *Apis mellifera.* How am I able to differentiate between the different races?

2B) What types of bees actually exist in the American market? What are the most popular and why?

3) What race of bees is most gentle? Which are most aggressive when handled in a colony of bees?

4) What type of bees is recommended for the Seattle area? We have much variation in moisture, humidity and temperature.

5) What other persons or organizations could I contact to assist me in the Seattle region in starting my own hive?

Jaime Garcia
Seattle, WA

Answer - 1) Honey grades are very dependent on what plant produced the nectar from which the honey bees foraged. One species of plant to another differs sometimes greatly in the sugars and minerals it contains, moisture, etc. etc. Then, add in differing flavors and quite a variety is available. The color of the comb also affects honey color–darker, old comb producing darker honey.

1A) Water White	*Extra White*	*White*	*Extra Light Amber*
Light Amber	*Amber*	*Dark Amber*	

2) There is so much variability in honey bees and locations that choosing a best race or strain is almost impossible. As a generality, they all have good and bad points.

2A) Generally, lighter colored U.S. bees are termed "Italian" and darker races are Carniolan, Caucasian, etc.

2B) Generally, the three recognized U.S. varieties are Italian, Carniolan and Caucasian. Of course, in the Southwest we also now have Africanized (scutellata).

2C) Midnite hybrid bees are almost totally black, but whether they are the ones you are seeing, is anyone's guess. They could be simply a Caucasian bee or a cross between two darker races.

3) Variable again - no firm pattern. Some bees are calm at certain times and aggressive at others. Darker U.S. races (Caucasian and Carniolan) have often been cited as gentler than Italian. The Midnite hybrid created by the late Dr. G. H. Cale, Jr., of Dadants was developed specifically for gentleness. It was advertised in the literature as "gentle as a kitten".

And, of course, Africanized bees are generally more defensive than other U.S. honey bee varieties.

4 & 5) Contact James Bach, State Apiarist, Wash. Dept. of Agric., Plant Services Branch, P.O. Box 42560, Olympia, WA 98504-2560, phone (360) 902-2068. Also, contact the Puget Sound Bee Assn., Frank Fitzpatrick, 18541 Marine View Dr. SW, Seattle, WA 98166.

Question - I am a beginner in beekeeping and have begun with just four hives. I am interested in the bees as a naturalist, and honey production is so far secondary to me. For this reason I've built a large indoor observation hive. It is just one frame wide, but 3 long and 7 high. I've carefully observed the 3/8" bee space, and the seven supers have narrow wood frames and large glass panels. About 95% of the bees will be visible at any given time in the hive. The bees will enter through the side of my house through the entrance which is T-shaped and equal in total size to a normal Langstroth hive.

Before I put my bees into the hive, I'd like to ask your advice about what race to choose for it. It's most important that they don't mind distractions and light coming into the hive, that they don't cover the glass with wax and propolis, and that their queen be happy to wander great distances in order to lay her eggs.

I'd also like to ask you if there isn't some sort of light I could use in the room which the bees would not perceive, but in which humans could see quite well?

Finally, can you tell me the optimum room temperature in which the bees could maintain their hive at 92° with the least difficulty?

Salz Jedlander
Lynwood, WA

Answer - It sounds as though you have constructed a very interesting observation hive. I wish you the best of luck with it.

As far as the type of honey bees to use in the observation hive, most any race will be fine. The Caucasian race is generally more docile than the Italian race. However, on the other hand, the Caucasian bees do propolize more heavily than do the Italians and for that reason you may wish to consider Italians to cut down on the propolization.

You also asked about the light distraction. When you are not observing the observation hive, it is a good idea to cover it with cloth or some other type of materi-

al that will shield it from the bright light and allow the bees to work normally as they do either in a man-made hive or a natural tree or cavity hive.

You asked about a special light that the bees could not see. If you use a red light such as those used in a dark room, you shouldn't have any problem with the bees being able to detect that. However, having a permanent red light as your only lighting source in your room may not be convenient either. This is something you'll have to decide on your own.

If you plan on keeping the observation hive year round, you will have to make some type of provision to make sure the bees have plenty of food for overwintering and you probably will need to be able to feed them syrup or honey.

As far as room temperature is concerned, you shouldn't worry about keeping the room too warm, in fact a normal room temperature of 60° to 75° F would probably be fine. The bees are very well adapted at keeping their cluster at the proper temperature despite the outside temperatures.

Question - In "The Classroom" July 1992 about transferring bees from an old hive body, you suggested putting a new body on top of the old hive body. My question is, when do you do this?

One other question from the same issue is about wax moth. In using paradichlorobenzene formulation, how long do you air out the combs before you put them back in the colonies?

Kenneth Dreese
Larned, KS

Answer - *The best time to put on an empty hive body with frames and foundation or drawn combs is early in spring. As the colony naturally wants to expand, they will utilize this space readily. Once the queen moves up into the new hive body, you can remove the old one and refurbish it.*

Your other question dealt with how long to "air out" supers that had been stored with PDB wax moth crystals. At least one day and a couple days would be better before putting the supers on the hive. They should take to it quite quickly if a nectar flow is on.

Question - Can I move my hive of bees during cold weather to a move favorable location?

Also, if I supplementally feed sugar syrup in spring, can I use Fumidil-B, a spring treatment for nosema?

Steven Rostetter
Oklahoma City, OK

Answer - *Yes, you can move your hive during winter if you have in mind one facet of the honey bee's wintering response. Honey bees in a colony start moving in together to form a cluster at around 47° F. They do this to maintain warmth for themselves,* *the queen and any brood that exists at that time. Because they are affected by cold resulting in slower movements and less strength, bumping, jarring, jiggling, etc., that may happen in a move, you may dislodge some of the cluster. They may fall to the bottom of the hive and if it is cold enough, not be able to regroup and rejoin the rest of the cluster. Many times they will freeze to death or starve to death when separated from the whole cluster. If this dislodged cluster contains the queen, you can foresee the future of the colony. All of that to say this – wait until the temperature is above 50° F and will be for a while, and be as careful as possible.*

If you are supplementally feeding your colonies, then the addition of Fumidil-B for the control of nosema is a prudent move. There are also some studies going on now that indicate that the feeding of Terramycin in the spring is also a good idea, but not as you would think for only AFB or EFB. Researchers have found that Varroa and Tracheal mites are either carrying bacteria that enter the honey bees when they bite and feed on its blood or that the wounds the mites leave allow bacteria to enter the weakened bee. Terramycin kills these bacteria and yields a healthier bee if the mite hasn't killed them or the colony yet. Final word on this isn't out yet, but it makes sense. We will wait and see.

REUSE OF DAMAGED COMBS

Question - I have several full depth frames of comb that were about 50/50 with pollen and honey. In the process of uncapping and extracting these frames, a lot of the comb was crushed with the pollen still in it. Can these frames be re-used? Will the bees clean the pollen out and fix the comb?

Don Ebrite
Villa Park, CA

Answer - The bees will repair and replace any damaged cells on the comb that they are given. Your combs are by no means a loss. Whether they will use or can use the pollen stored in the damaged sections I can't answer. It is a 50/50 proposition if they will want to use that pollen. Regardless, put your frames back in this spring and the bees will use them.

Question - These are some of the questions that I, with years of experience in bee-keeping, do not know the answer. Can you help me, please?

1) Do worker bees move eggs from worker bee cells to queen cells or to other worker bee cells?

2) Do worker bees move honey from central brood combs to less crowded cells, i.e., outside frames or to supers? If so, is it always cured honey?

3) Does the queen bee feed herself? Does the drone feed himself?

4) Does the queen bee lay eggs in a pattern when she has a choice, i.e. circular / spiral? If so, in what direction, and would this direction be reversed in the southern hemisphere?

5) Why is it that drone cells on a frame are usually found on the opposite side in a back-to-back position?

6) When I remove pollen pellets from pollen traps there are plant seeds in with the pollen, but not attached to the pellets. Some of the seeds are quite large. There is no shortage of pollen when this occurs. Why are the seeds collected?

7) I have read that stingless bees have a "nasty habit". What is it? I observed stingless bees in Costa Rica; they got in my hair, but I didn't consider them to be offensive.

Fred Fulton
Montgomery, AL

Answer - Gee, Fred, you sure come up with some good questions. Some of my answers may not be as complete as I or you would like, but I'll give it a go.

1) There has been a rumor and anecdotal information for hundreds of years that worker bees move eggs from one cell to another. These rumors have many times been based on sightings of workers with eggs in their mandibles. No one has ever seen a worker removing an egg from one cell and successfully placing it in another. Workers will remove and eat eggs for a variety of reasons, starvation in the hive, imperfect eggs for one reason or another, etc.

2) When honey has been cured and capped workers do not usually move it from one location to another. Sometimes uncured or unripened nectar will be moved initially from cell to cell, but not often. Honey bees at this level are pretty efficient and do not expend any more energy than necessary in this effort.

3) A queen can feed herself honey or nectar for survival purposes, but to be able to lay eggs, she is fed a mixture of brood food and honey by attendants. Drones after seven days of age feed themselves.

4) She generally isn't haphazard in laying so some pattern is probably used, probably determined by pheromones and available space.

5) Honey bees, being the structural engineers that they are, know that for structural integrity of the comb that this alignment is needed.

6) Honey bees collect all sorts of interesting things, seeds included. No one seems to know why items that can't be used readily would be collected but, they are. It's instinctive to gather pollen-like material, even if the real thing isn't present.

7) There are several species of stingless bees in Central and South America. Their two major forms of defense are biting with their mouth parts and if that doesn't work some species spit formic acid which if gotten in eyes or mouth is very uncomfortable.

HIVE DEBRIS

Question - The weather was almost 50 degrees, so I checked my bees. I took off the covers and saw the top of the cluster about 2/3 of the way down the middle frame of the top hive body with the rest filled with honey. I assumed they were okay. There were 20-25 dead bees out front and clogging the small hive opening so I cleaned them away and opened the hive up all the way across since it was the very narrow side of the reversible bottom and since I was hoping that would get rid of some of the moisture that had collected on the inside of the outer cover. As I scooped out the front of the opening to get dead bees, I also got something that looked like cornmeal or pieces of pollen that was coating everything. I seemed to remember reading somewhere that this was not a good sign. The particles did not stick together so I don't think it could have been wax and it was the wrong color of yellow. What is your opinion of what I have described in my hive? I am also considering getting another package or two of bees this spring and wondered if I could successfully install a 3-pound package in a 4-frame nucleus colony on foundation until they build up. Thanks for your help.

Mike Fillenwarth
Indianapolis, IN

Answer - Your first question dealt with hive debris at this time of year. You have to remember that for the most part all hive activity is continuing in winter within the hive. That means that cappings are being removed to gain access to honey and pollen, some hive/comb repair may be going on, comb cleaning to prepare for the queen to start laying again, some brood may be developing or just started or just removed etc., etc. All these activities generate some waste and debris which is simply dropped to the bottom of the hive because the bees cannot consistently get outside to carry this material away. The same waste is generated at other times of the year, but is carried outside the hive by the house-cleaning contingent of honey bees. So, don't worry it's just trash the bees have generated and haven't removed.

Go ahead and install the three-pound package on a small nuc and feed them syrup until everything is drawn out and brood developing. When you do your spring colony work, be sure to scrape the bottom boards clean of dead bees and other debris. This is also a good time to check the condition of bottom boards and replace them as necessary. Have some extras with you for this purpose.

Question - I am writing to you for the first time to ask you a few questions about honey and honey bees. It would be very kind of you to answer them in the next issue.

1) Why is the color (clarity) of honey one of the important factors when grading honey, and does grade mean quality?

2) Should we put inner covers on a hive everytime even when it is too hot or when the humidity is high? I think that a telescoping top cover with two small windows in two sides will make air currents to ventilate the hive.

3) Do you think there is a kind of honey anywhere which can help a specific disease like cancer or skin disease.

Abdullah Alkhudairy
Riyhad, Saudia Arabia

Answer - *Your first question dealt with how or if honey clarity is related to honey quality. When I think of honey clarity, I'm not so much concerned with color as I am with its cleanliness. Clarity means the absence of any extraneous matter suspended in the honey. Bits of wax or pollen, lint from filtering cloths, bee parts (legs, antennae, etc.) even air bubbles reduce the clarity of honey. Even though these things most likely will not hurt the consumer if eaten, they give the impression to most consumers that the product was hastily and cheaply prepared. It is the honey processor's responsibility to make sure that the honey offered to the consumer is as pure as when the bees put it in the comb.*

Color is largely a matter of taste. Lighter honeys are usually milder to taste and evidently a majority of consumers prefer mild honeys over darker, stronger flavored honeys. Therefore, they fetch a higher price in the marketplace.

The most important reason to use an inner cover with telescoping lids is that when the bees glue the inner cover down, the telescoping lid can still be easily removed. If there is no inner cover, the bees will glue down the telescoping top, making its removal very difficult without disturbing the bees greatly. The inner cover's value is negligible after that. I would agree with you that establishing an upper entrance in your colonies would be a great benefit for heat and humidity control.

When it comes to using honey, propolis, pollen, bee venom, or any other hive product for specific medical reasons, I do not think there has been enough research or investigation completed. The medical community has been slow to follow up on reports of the therapeutic benefits of honey bee-produced products. I do not think that there has been enough documented honey and bee product research. We definitely need more rigorous research by the medical community. Anecdotal literature certainly lends support to the belief that these products are useful.

Question - 1) If honey is supposed to be nectar, then feeding bees sugar and water would not be nectar, right?

2) If feeding sugar and water, then would it be against the law to have "Pure Honey" on your label?

3) When bees swarm, what is the distance they will travel before settling in a new place?

4) When eating comb honey, is there any food value there?

5) One time there was a black snake among my bee hives - sometimes he would just look at me, other times he would crawl under the hive. He was there about two months. I was thinking about taking him to the barn. One day he laid dead about four feet in front of the hives. He was not harmed in any way. What was his purpose in being there and what could have happened to him?

Steve Chiolerio
Marceline, MO

Answer - *1) Right, I suppose that there may be some unscrupulous beekeepers around who may feed sugar syrup to the bees past the time when it is usually fed in early spring to counter low food supply or stimulate brood rearing, but there are eas-ier ways to adulterate honey than by feeding sugar or corn syrup to honey bees.*

2) Yes, it would be against the law. Ask the Chinese about this. They have had numerous loads refused in several countries because of adulterated honey.

3) Generally as close as possible for European bees. In most locations there are hollow trees or spaces in a wall of a house or an attic or anything with some space and an entrance. If a European swarm travels over a mile, it would be unusu-al.

The Africanized bees, on the other hand, will send out reproductive swarms and will abscond themselves and travel many miles to find a new location.

4) There is obviously food value in the honey in the comb, but if you are ask-ing about the wax of the comb, then the answer is no. Beeswax is indigestible and has no food value except for the bulk it provides.

5) Snakes like secluded safe places and under a bee hive is about as good as it gets for a snake. I've seen many snakes under hives and when moving colonies I always check just because they always startle and scare me. What the snake could have died of I couldn't guess.

Question - I wanted to kill off a colony of diseased bees, but I didn't want to burn them. I was going to burn the frames, combs, etc., but later I thought it too inhumane to burn the bees. This is what I did and it didn't work.

I thought I could kill the bees by replacing the oxygen in the hive with nitro-gen. I rented a small cylinder of nitrogen from a welding supply company and hooked hoses up to it. I then taped up the cracks in the hive sides, bottom and entrance with duct tape. I stuck the hose in the top and opened the value, letting the nitrogen in slowly and left it like this all night. The next morning I checked the nitro-gen tank and half of it had been used, but the bees were not dead. I thought that if you replaced the oxygen with nitrogen, the bees would die. Don't bees need oxygen? Can they hold their breath for hours? What's going on?

Walt Wright
Elkton, TN

Answer - *Yes, bees do need oxygen. Just like any other breathing insect, reptile, mammal, fish or bird, they need oxygen to be able to carry on and complete the com-plicated chemical process and reaction going on in their body all the time.*

No, bees cannot hold their breath for hours. Here are a couple of things I do know:

1) If you had eliminated 100% of the oxygen, the bees would have died.

2) The hive could not have been completely sealed to eliminate the nitrogen from getting out and oxygen from getting in.

3) Bees when at rest use an incredibly small amount of oxygen.

4) Cool these bees and lower their metabolism and this incredibly small amount of oxygen used gets even smaller.

5) Nitrogen as a compressed gas, like you had, when vented into the atmosphere absorbs heat. It has a cooling effect going from high pressure to low pressure.

6) A higher concentration of nitrogen in the air mixture the bees were breathing has an almost anesthetic effect on them; further calming and lowering their metabolism and thus their need for oxygen.

*Here's what I think happened: You put the hose from the nitrogen tank under the hive top into your frames. The hive, not being completely sealed, allowed the nitrogen to seep out and, of course, oxygen to get in. The flow rate of the nitrogen was not great enough to displace all of the air and its oxygen content. The nitrogen flowing into the hive had a cooling, calming effect on the bees, making their need for oxygen greatly reduced. As a result, no dead bees! *Any readers out there with a different or better scenario based on the information available, please drop me a note. Thanks.*

VENTILATION

Question - Every afternoon when it is hot, the bees form clusters outside the hive. Do they need more ventilation? Later when it gets cooler they go back into the hive.

Answer - The bees likely hang on the outside because the colony is heavily populated and overcrowded. They need shade, ventilation and some attention or they will solve the situation themselves by swarming. First, completely remove the entrance block to give the bees full access to the hive entrance and also to aid in ventilation. Then let's look at your supering to see if the bees need another super added. This would ease the overcrowded situation and give those bees something to do in drawing the foundation in the frames of the new super. If more ventilation is required, the inner cover could be moved back from the front edge about a quarter of an inch, giving more air space. The supers could also be staggered slightly front and back about bee space width to further increase the movement of air through the hive. If additional ventilation is needed, you might raise the hive itself off the bottom board by inserting small blocks of wood at the corners. All of these techniques will aid the bees in keeping the temperature of the interior of the hive near normal, but be careful about leaving the hive in such condition for long periods of time as you will have opened up more spaces than the bees can defend and you may encourage robbing. You might also consider a light shade such as placing the hives in the fringe area of the shade of a tree with light open branches.

HOUSE APIARY

Question - Because my urban yard is surrounded by neighbors and pets and is very small, I would like to begin my beekeeping hobby by placing two hives on the second floor of the house. I would certainly appreciate any thought that you may have to offer concerning potential problems etc.

Diane Dodson
Richmond, VA

Answer - Your idea is a good one and will certainly reduce some potential problems with your neighbors. A secluded out-of-sight location on the second floor of your house obviously will not trigger immediate unsubstantiated fear by your neighbors and it will force the bees to fly freely unobserved over the heads of everyone. Assuming that this is the attic of your house, be sure that there is adequate ventilation, especially in summer and adequate room for you to open and inspect your colonies, add supers, etc. Beekeepers have been keeping honey bees in buildings, for hundreds of years in Europe, so I'm sure you'll be successful.

ENCOURAGING HONEY BEE POPULATIONS

Question - In 1991 I purchased a tract of land in Tennessee which I intend to convert to a medium-size cow-calf operation in a few years. It is currently rented to a neighbor who runs a few head of cattle on it. Since there is very little clover on the property, I am trying to spread it through over-seeding. To help pollinate the clover, I would like to encourage a bee population - I have yet to see a single bee on the property. I am unable to locate anyone in that area who manages beehives and, even if I did, I doubt they would be interested in the one or two hives I might require until the clover becomes widespread.

Is it possible to make a "hollow tree" equivalent by using a white plastic five gallon container hung from a tree limb? What I had in mind was putting an opening under the bottom and air holes under the lid rim for ventilation. If this is feasible, what could be used for inside comb attaching material? How could I initially stock a colony and then start others? What size openings would be needed to keep out bats?

Ken Scharabok
Dayton, OH

Answer - I would encourage you to contact Mr. Thomas C. Hart, Tennessee Dept. of Agriculture, P.O. Box 40627, Melrose Station, Nashville, TN 37204, Phone 615-360-0130, who is the Chief Apiary Inspector in the State of Tennessee. He would be able to help you locate a beekeeper who may want to locate honey bees on your property.

I would also discourage you from trying to establish honey bees that are permanently in a hive with non-movable frames. This colony would be unable to be inspected for disease or africanization and may be a reservoir for disease and unmanageable honey bees. Mr. Hart can tell you about beginner beekeeping programs in the state and help you become a beekeeper who is an asset to beekeepers everywhere and to neighbors in your area.

WHAT IS A SPLIT?

Question - Lately, I have seen people who had splits for sale. My question is: What exactly is a split?

Rudy Haynes
Holyoke, CO

Answer - *A split is when a colony is actually divided or split into one or more additional units. Most splits are made when four or more frames of brood, honey and bees are removed from a colony and then are given a queen. This creates a small independent colony which then can be put into standard equipment at some point and allowed to expand. The advantage over a package of bees is that there is drawn comb available in a split, along with brood ready to emerge, and a queen that is capable of laying immediately. This allows it to develop more quickly.*

REMOVAL OF HONEY BEES FROM A HOUSE

Question - We have bees in the peak of our very large three-story old home. They've been there for over a dozen years. Do you know how we can get rid of them? Can you recommend someone who may help rid up of this problem?

Paul Carroll
Waterville, NY

Answer - *I hope my reply doesn't sound too sarcastic, but if you have had honey bees in the wall or roof space of your home for over 12 years, you do have a problem. As you can well imagine, after 12 years a colony probably has expanded its nesting area extensively over this period of time. Because of this, getting the colony to move out willingly or forcibly will be complicated. Other considerations would be surplus honey still stored in the combs and what to do with it – can it be removed? What about any electrical wiring? Have the bees chewed on this for 12 years making it a hazard, etc. etc.? My suggestion would be to call our branch in Waverly, NY (607) 565-2860 and inquire if they know of any beekeepers in your area who may be willing to take on this project. It can be done, but it will not be easy, quick, or cheap, so bear that in mind. You also may need to consult with a carpenter if areas of the house have to be dismantled in order to physically reach and remove the bees and their combs. Be sure that all holes are plugged after removing the bees, so re-infestation will not occur.*

BEES IN A HOUSE

Question - I've heard that hummingbirds beat their wings very quickly to maintain flight. What is the wing beat frequency of the honey bee?

Alec Gifford
Louisiana

Answer - *That's not a figure that I could pull off the top of my head so I went to Eva Crane's book <u>Bees and Beekeeping.</u> "The wing-beat frequency in flight is quoted as 235 or 250/second for European workers (Dade 1962); the frequency is higher at greater speeds. It is slightly but noticeably higher for tropical African workers, whose wings are smaller. H.T. Kerr and Buchanan (1987) measured the dominant frequency as 210/s for European bees and 270/s for Africanized bees. In Punjab, India, for workers and drones, respectively, frequencies were 235/s and 225/s for <u>A. mellifera</u>, and 306/s and 283/s for <u>A. cerana</u> (Goyal & Atwal. 1977)."*

Question - What is ear candling? And, what does it have to do with beekeeping?

Answer - *Ear candling is one of the most unique (bizarre) looking techniques to give relief from ear wax build up, earaches, etc. I'll let you read the following reprint from the British Columbia <u>Bee Biz</u> Newsletter for the complete description.*

The only connection ear candling has to beekeeping is that beeswax may be used to make the device used in this procedure.

P.S. Not being one to shirk my duty as a purveyor of cutting edge beekeeping information I tried ear candling. It seems to work as advertised, but I'm glad no one took my picture while doing it. Remember, I am a professional. Do not try this at home by yourself.

EAR CANDLING

For those beekeepers who think they've seen it all, we look at an intriguing application of beeswax candles.

The fast-growing practice of ear candling, or coning, whichever term you prefer, is becoming commonplace in Canada. The technique is not new. It has a history of hundreds of years in such places as India, China, Egypt and in Native American healing arts. Patients are vocal in their satisfaction. Relief from wax buildup, headaches, ear infections and sinus irritation, itchy ears, and even some problems where a hearing aid has been recommended are reported. While practitioners are quick to point those with medical problems to see a doctor, and to discourage people from practicing ear candling on themselves (no reason given in the literature seen to date) this simple, natural process must be placed among the most practical of alternative healing methods. Historically, a number of variations have been used. Rolled up newspaper dipped in wax was used during the Depression. Native Americans used mullein stalks and hollowed corn husks. Anything like a hollow rod was used, as long as it would burn.

The candles, or cones, used today are made either by dipping a cotton strip in beeswax and shaping it around a core, or by first wrapping the cotton onto the *core and then dipping it into the wax. The result looks very much like an ordinary taper, but is hollow. The amount of wax relative to cotton is important...too much wax will drip and run. Some candles have herbs added to them, which are said to aid in the healing process. A variety of sizes are used, each practitioner finding his or her own preference after trying the wares of several producers. Prices range anywhere from $1.50 each in large, wholesale quantities, to $4.00. A number of beekeepers have become involved in the business, and in some areas, competition is becoming keen.*

The ear candling process is simple. The candle/cone is pushed through an aluminum foil pie plate (to keep ash and any drips out of the patient's ear). When the candle is lit, a vortex (suction) is created inside the hollow tube, which draws up ear wax and other debris. Some of the swirling smoke also enters the ear canal, warming, smoothing and relaxing the inner walls and loosening the wax that has formed there. The process takes an hour or so, and each session requires two or three candles per ear. Practitioners point out that it may take anywhere from one to seven sessions to totally clear the ear canal from wax and infection residue.

After a candling, the ears must be protected from cold for 3 hours and from wet for 24 hours. Putting cotton into them is advisable when going outdoors or taking a shower. Ear wax naturally replenishes itself within 24 hours. The results are impressive. And working with practitioners can be a lucrative sideline or winter project for beekeepers with the time and patience to develop a quality product.

Apitherapy and ear candling are only two of the constantly growing number of fields concerned with holistic/natural/alternative medicine that can make use of beekeeping products. As pressures mount on health care systems, and people demand more preventive, less invasive medical practices, more health professionals are looking at traditional folk-medicine and finding a lot more than they expected. This is a trend that can be expected to continue for the foreseeable future. Stay tuned for further developments. Information supplied by practitioner Jasmine Espert, Jasmine's Esthetique.

Question - Hi, my name is Katy. I am 10 years old. My dad is a hobby beekeeper. I have two questions for you. Number one, about how many drone bees get kicked out during the winter? My second question is about how many pounds of honey do the girl bees collect a day? I hope you'll be able to answer my questions.

Katy Wareham
Moses Lake, WA

Answer - Drones are kind of an insurance policy for a colony. Since they do not do any "work" to provide food, defense or construction to the hive, they must have some value or they wouldn't exist. That value is to provide the genetic mixture needed, after mating with a virgin queen, for the fertilization of eggs and the continuance of the species. If there were no other drones available, then these home grown drones would provide the mating sequence and its results. In other words, they are ready to be dads if needed.

Where the winters are harsh as here in the Midwest, drones are not of any value in the winter where development, rearing or flight of bees is not possible and thus drones could not provide any benefit. So in the wisdom of the workers at this time of year (fall) the drones are pulled out, harassed and not welcomed anymore in the hive. But in warmer parts of our country drones may be tolerated year round.

Your second question asked how much nectar/honey do bees collect in a day. This is highly variable. It is based on population, hours of daylight, wind velocity, temperature, predators, diseases and of course, type and volume of nectar available. I've seen from 0 pounds to around 5 pounds of nectar brought in per day in my own colonies. I'm sure that in other parts of the country it may be more.

Question - I live in an area that up until a few years ago, was classified as rural. That's when people from other than rural backgrounds started moving in. At first, everything was all right, then the complaints started coming in about local farms and the smells and noises that came from farming. Now the complaints have broadened to include honey bees, my honey bees! There is talk about a ban on honey bees in our community. What have other beekeepers done in this situation?

Todd Karawaski
Elgin, IL

Answer - *Hindsight is always 20/20. You should have seen this one coming. When anyone – me, you, your nonrural neighbors come into a new situation, we react sometimes in an irrational manner because we are ignorant. Ignorance means we don't have complete information and use the incomplete information and prejudices we have to make unsound decisions.*

#1 - You need to start an education process in your community about the importance of honey bees – videos, slide presentations, talks for business, service and school groups. This will show the worth of honey bees and mitigate some of the bad feelings some may have about them. Your goal is not to turn everyone into a honey bee lover, just a honey bee appreciator – a creature, that deserves to live in your community.

#2 - If push comes to shove and your local legislative body is considering banning honey bees, submit the Sample Ordinance below for review and negotiation. This ordinance allows beekeeping within guidelines generally acceptable to beekeepers and nonbeekeepers.

SECTION 1 - <u>Location of Bee Hives and Other Enclosures</u>. It shall be unlawful for any person to locate, construct, reconstruct, alter, maintain or use on any lot or parcel of land within the corporate limits, any hives or other enclosures for the purpose of keeping any bees or other such insects unless every part of such hive or enclosure is located at least seventy-five (75) feet from a dwelling located on the adjoining property.

SECTION 2 - <u>Number of Hives (Colonies of Bees)</u> Regulated. On lot sizes of 15,000 sq. ft. or less no more than 4 hives (colonies of bees) will be permitted. The *hives shall be no closer than 15 ft. from any property line. On lots larger than 15,000 sq. ft. additional hives will be permitted on the basis of one (1) hive for each 5,000 sq. ft. in excess of 15,000 sq. ft.*

SECTION 3 - *Type of Bees.* *This ordinance shall pertain only to honey bees maintained in movable frame hives and it does not authorize the presence of hives with non-movable frames or feral honey bee colonies (honey-bees in trees, sides of houses, etc.).*

SECTION 4 - *Restrictions on Manipulating Bees.* *The hives (colonies) of bees may not be manipulated between the hours of sunset and sunrise unless the hives are being moved to or from another location.*

SECTION 5 - *Penalty.* *The violation of any provision of this ordinance shall constitute a misdemeanor punishable upon conviction by a fine not exceeding fifty ($50) dollars, or imprisonment not exceeding thirty (30) days, provided, that each day that a violation exists or continues to exist shall constitute a separate offense.*

#3 - Be a good and patient neighbor. If you proceed in an intelligent manner, you'll get along much better than if tempers flare and relationships are broken. Don't burn any bridges behind you. A little honey given as gifts sweetens many temperamental dispositions.

Question - Now that the searing heat we had this summer is but a distant memory, can you tell me what I should do to protect my bees from high temperatures? I was very worried about them when heat and humidity reached dangerous levels this summer.

Michael Gram
Broken Bow, OK

Answer - We here in West Central, Illinois had one of the hottest and wettest summers that I can remember. We had temperatures over 100°F and humidity in the 80-90% bracket. It was miserable for humans, livestock and bees. Only plants seemed to appreciate it.

There is not a whole lot you can do for the bees that most beekeepers don't already do. The most important is to obviously keep the bees out of constant direct sunlight if possible. Paint your hives a light color to reflect as much sunlight/heat as you can. High humidity for all animals is the main problem. Give your bees some upper ventilation that will allow air to flow through the hive. Just cracking the top open a 1/4 to 1/2 inch with a small stick can remarkably cool off a hive that is too hot. If not available naturally, provide a water source for the bees to collect and use in cooling.

Do these things and with the help of your bees you should be O.K.

Question - 1) Can comb or extracted honey be kept in the freezer and not granulate indefinitely?

2) What protection is available for wax moth protection for supers?

3) How do the large packers keep honey from granulating on store shelves?

4) Could you send me a monthly check for $1,000.00 for life so I could afford to keep more bees? Yes! I'd settle for $750.00 each for month for life.

Charles Miller
Pekin, IN

Answer - 1) Indefinitely? In the kind of freezers and the temperatures they operate at for home use, probably not indefinitely. But who wants to keep honey indefinitely in a freezer anyway? Home freezers will keep most honey granulation for a year or so at 0°F.

2) There is no chemical control available that is registered for wax moth control in honey supers. If you will review the last year to year and a half of the Classroom, this question has been asked several times. You'll find more complete answers there. Suffice to say, these answers contain information on carbon dioxide treating, use of lights, etc.

3) Large packers strain and filter the honey to remove particles which hasten granulation, then they flash heat the honey and then quickly lower the temperature back to under 100°F. The heating melts away crystals of sugar in the honey which starts the granulation process and also may kill yeasts that start fermentation. The temperature is lowered quickly to stop the loss of flavor and darkening of honey that high extended temperature can cause.

4) I think you are underestimating the amount of money needed to keep bees.

Question - In 1978 Drs. Harris and Stone, in the book "All About Allergies," page 245, stated that apiarist asthma is caused by inhalation of the emissions from honey bees, associated with bee pollen.

In the subsequent years I have heard nothing about this malady. Have you or your readers? In the fall I have noticed a musty pollen odor in and near bee hives. Maybe I can lay the cause of my postnasal drip all these years on apiarist asthma!

Fred Fulton
Montgomery, AL

Answer - I've heard the same thing, but as far as I know there has been no formal research about the link between the smells originating from a bee hive and beekeeper asthma or any other respiratory condition.

The only study I've ever seen is one done about 10-15 years ago which showed that the beekeeper's family was something like 60% more likely to have an allergic reaction to bee stings than the public at large. These family members were not involved in the day-to-day work of beekeeping, but they were exposed to dried venom on the beekeeper husband/father or his clothes and gloves which sensitized them to stings.

So whether it is pollen, dried venom, propolis dust or molds, mildews or fungus, I think that there may be some connection between these specific ailments of beekeepers and their families. but until someone spends some money and time on formal research, it's just my opinion.

The "musty pollen odor" you smelled in the fall is probably the smell of fall honey being ripened by the bees. Oftentimes, fall honey has a strong odor as the excess moisture is removed by the bees fanning their wings. Smartweed, aster, goldenrod and spanish needle are the fall sources most often mentioned as causing "musty smells" from the hive in the late summer or early fall.

Question - With fall approaching, I have several supers of uncapped honey. Should I leave them on the hives for next year, or will the honey spoil?

Also, I heard that if placed in a ziplock bag and stored in a freezer, Apistan will last for approximately five years. Is this a true statement?

Colby Mullen
Williamsville, VA

Answer - *Hopefully, in the next few weeks the bees will have capped those frames that are uncapped as of now. If the honey has a moisture level of above 18%, whether it is in the hive or in your house, it will start to ferment if the temperature is warm enough. If you have the freezer space, you could freeze these frames to save for later or even extract the honey and freeze it for feeding back to the bees or to be used quickly to cook with. Give the bees a little more time, they may still cap it.*

Apistan, if kept in a sealed container in a cool, dry, dark place, will last a number of years. I think under these conditions, five years is easily attainable.

Question - I just returned from our regional beekeepers' association meeting. Our state apiarist was there and out in the hallway he was talking about more growers using "pollen" bees. I didn't overhear a whole lot more before I went to the next talk in the auditorium. What are "pollen" bees? Are they a specially bred honey bee or a different kind of bee?

Oscar Strasser
Oakland, MD

Answer - *Did you know that of the 20,000 known bee species that only seven are honey bees (genus Apis). The rest are "pollen" bees also called wild or solitary bees.*

The short answer is that "pollen" bees are bees that do not store honey. They are very gentle, rarely sting, live in small groups and collect pollen like crazy.

They certainly don't fit the description of bees as we think of them and they are not wasps or hornets. Instead, they go by names like digger, sweat, bumble, horn-faced, carpenter, leafcutter, orchard and shaggy fuzzyfoot.

These types of bees are incredible pollinators in comparison to the honey bees we know the best. As an example, a single hornfaced bee can visit 15 flowers a minute, setting 2,450 apples in a day - compared to the 50 flowers set in a honey bee's day. That's why you need thousands and thousands of honey bees in many colonies to do the same pollination as hundreds of these other bees. So, if you are looking only at pollination, honey bees may not be your best choice. That's why you overheard the discussion you did in the hall. The drawback at this time is learning to rear this type of bee so that it is practical to use them as pollinators in the future. This is what researchers are working on now.

If you would like more information on these bees and specific "pollen" bees suited to your area, contact the USDA-ARS researchers below. I'm sure they would be glad to help.

Dr. Suzanne Batra
USDA-ARS Bee Research Lab
Bldg. 476, BARC-EAST
10300 Baltimore Avenue
Beltsville, MD 20705-2350
Ph. (301) 504-8205

Dr. Stephen Buchmann
USDA-ARS Carl Hayden Bee Res. Center
2000 R. Allen Road
Tucson, AZ 85719
Ph. (602) 670-6481

Dr. Philip Torchio
USDA-ARS Pollinating Insect Biology Res. Unit
Natural Resources Biology Bldg.
Utah State University
Logan, Utah 84322
Ph. (801) 797-2520

Question - I'm new to the beekeeping industry and enjoy it as a hobby. I only have two hives at the moment, but I plan on increasing in the future. As you can imagine, I have several questions concerning my operation. I do subscribe to *ABJ* and that has helped a lot. I have located a person here who has raised bees, but I have bothered him so much it makes me feel uncomfortable. I was wondering if there was someone on your staff who could answer my questions. I know some of these are real basic, but being new to the bee raising I bring a very ignorant approach. I have gone to the library and read all the books they have, and that has helped also. Please help if you can. Thanks.

On each hive I have: 2 deeps for brood chambers, then 1 shallow followed by 1 deep used for honey storage. (1) When do I put a queen excluder on and for how long do I keep it on, and will putting it on cut down on my honey production? (2) When do I start extracting honey (I have a small plastic extractor here at home with the bees)? (3) When do I divide a hive or in fact should I even divide it? (4) How can I control the size of my colony? If I only wanted two colonies how do I stop it from growing? (5) When should I requeen? (6) Can I let the bees requeen themselves? What are the pros and cons of doing this? (7) Do I put my Terramycin on the top of the frames of just my brood chambers or do I put it on the supers also? (8) The tops that I am using have a round hole right in the middle of them. Doesn't this bother the bees when it is cold at night? (9) I want to check each of my hive boxes from the first brood chamber to the top super. I know I should do this when it is warm, but can I take it apart without causing the bees to get mad and leave? I mean I want to check out each frame in my hive and try to locate the queen and just look at how things are going. As you can tell, I'm very excited about my bees. (10) Do you know where I can get small decals advertising honey and bees that I can put on my vehicles? (11) We get a fair amount of snow here and I have heard of two different ways to winter the bees. One is just leave them alone and they will take care of themselves (sounds cruel to me). Two is put straw around the base with a tarp on top and this will do. What do you recommend? (12) How often can I open my hives and look at them without causing a problem? (13) Do you know of any beekeepers in my area who I could talk to? (14) I know I will have more questions in the future as I go through the different stages of my operation, such as extracting, storing, etc. Can I write when I have more questions? (15) I would like to create a watering hole for my bees, do you have any ideas of what kind of setup I should create?

Mark Gosswiller
Boise, ID

Answer - Thank you for your questions. I appreciate your enthusiasm. You'll do fine. (1) The reason a queen excluder is used is to keep the queen out of your honey supers, so she doesn't lay in them and raise brood. Put it on when you put your drawn supers on. Don't use the excluder, however, until the bees have drawn out combs in your supers. They are less likely to cross the excluder to work foundation. Honey production is about the same. (2) You may start extracting any time after 95% of all the cells are capped on your comb in the honey supers. (3) If you want to divide for increase in colony numbers, early spring is good and with supplemental feeding they may build up enough to still produce a honey crop. If you don't want or need a honey crop, any time up until the middle of summer will work. (4) Don't divide – don't collect swarms. However, be sure to requeen at least every two years if needed and put your supers on early to accommodate your big colonies. (5) Yearly is optimum, every two years for sure. (6) Yes, this is what happens when they prepare to swarm. Pros are it is a cheap way to requeen. Cons are the quality of the queen and her offspring is quite variable in all respects, from excellent to terrible. (7) Terramycin is used to treat diseases of the brood specifically, so use it only on and in your brood chamber. You do not want possible antibiotic residue in your honey, so stop using it several weeks before adding your supers. (8) Not any more than all the air leaks in my or your house. (9) Fire up your smoker, so it produces a cool white smoke on a sunny warm day and inspect every nook and cranny. That's how you learn. (10) I hope you will join shortly or already have joined your local beekeepers' association as they can help you with this. (11) Snow is an excellent insulator. When I lived in Michigan we always hoped the snow would cover the hives as these survived the best. If you want, you can do any of the things listed. Overall, it really doesn't matter as long as the bees are dry and have enough stored honey available to them. (12) As a beginner, every week or two is all right. That's how you learn. (13) Mr. Michael Cooper at the Idaho Department of Agriculture in Boise at (208) 334-

2986 is the Idaho State Apiarist and he can put you in touch with any beekeeper in the state. Your state association contact is Jim Ellis in Emmett, ID. His phone number is (208) 365-2732. I suggest you contact him, too. (14) At any time. (15) Many beekeepers in the arid west use buckets or barrels filled with water that have small blocks of wood, etc., for the bees to land on. The bees use these as landing boards and fill up on water. More elaborate watering systems are available, including in-hive watering devices. However, for a small beekeeper they are not really necessary. Sometimes simple is best. If your bees are located within a half to three-quarters of a mile from a good farm pond, your problem is already solved. Streams that don't run dry are also good sources of water for your bees.

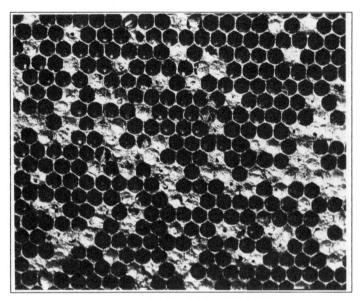

WHAT IS WRONG HERE?
(ANSWER PROBABLY AFB OR EFB)

Question - I have heard that an average hive during the honey season holds approximately 80 to 100 thousand bees. My question is this: During the winter days what does the population of a hive this size dwindle down to assuming that this is a healthy hive? And what are the factors that cause such a decrease? If, in fact, there is one.

Moss Gosswiller
Boise, ID

Answer - *Your main question was, "What does the population of a hive dwindle to in winter?" From the literature that I reviewed, the optimum size of an unrestricted colony in early fall should be approximately 30,000 young bees. The population can consume 15-20 pounds of honey and lose 3-5,000 individuals before egg laying* restarts. *Remember that this is a scientific optimum and individual hives will vary sometimes widely from this figure. Some colonies may have significantly more bees and eat up all the stored honey requiring feeding to survive and some may be smaller than the above and not be able to maintain enough warmth and die during winter, regardless of how much food they have.*

Bees are affected by day length. Physiological responses start as a result. This is no different from any other animal that uses clues in its environment to prepare for winter. In early fall as day length lessens and nectar supplies change, the worker bees raised are physiologically different from the workers raised in spring or summer. These bees have fat deposits on their bodies which allow them to live longer, than the 6-8 weeks in summer, and survive cold as they generate heat for the cluster which forms in response to cold. These bees dominate the colony as brood rearing finally stops in late fall or early winter and the "summer bees" finally die off as their life expectancy is reached. Hopefully, your colony now has 30,000 or so of these "winter" bees, plenty of food, a good location out of the wind, etc., no disease, no mites. If so, they may survive the winter and build rapidly in spring to collect a bounty of honey.

Remember, the above description is for the cold snowy north. The scenario is different in the southern states from east to west and some coastal areas in the Northwest. Honey bees are incredibly adaptable and have survival mechanisms which allow them to exist almost everywhere. That's why they are such remarkable organisms.

Question - I would like to offer advice to your reader about sugar spoiling in the winter.

With every gallon of sugar syrup we mix in a few crystals of (thymol). This is crystallized oil of thyme. Is it readily available in the USA? We use it if we feed late and the bees do not have time to cap it. It stops the syrup from fermenting. I had a gallon of sugar syrup in a garden shed all summer and it was okay to use next autumn.

Jean Moxley
Kent, England

Answer - *Thank you so much for your input on your technique of stopping the spoilage of unused sugar syrup.*

The use of the herb thyme and the oil in its concentrated form certainly seems like a natural method which should be looked at more closely. You guessed at the problem of use of Thymol here in the U.S.A., it's not readily available. Like many other things, people use what they have ready access to and Thymol unfortunately isn't one of them.

Question - I heard that stingless (split stinger) Africanized honey bees had been developed in Brazil, but had not been released for beekeeper use for I don't know what reason. If this story is true, we need this bee for American use.

Alexander Alt
Santa Barbara, CA

Answer - I also remember reading or hearing about a honey bee that was found to have mutated to the degree that its stinger could not penetrate anything. It seems there were two major problems with this bee. One, this mutation could not be passed on consistently to additional queens and drones for mating purposes and the bee-keepers in South America were very disinterested in a honey bee which couldn't defend itself against predators including man.

These two reasons killed any additional research in the project.

Question - Many beekeeping supply manufacturers have recently included swarm traps as part of their inventories. If I were to purchase one and set it up properly in an area where I have seen bees collecting nectar and pollen, do you know what my chances of obtaining success would be? I understand that both Tracheal and Varroa mites have had a large influence on reducing the number of the feral colonies. Do you know the current distribution of feral colonies per square mile in the U.S.? Will these mites have a large influence on whether or not I will find a swarm?

Dominic Dallago
Pottsville, PA

Answer - Swarm traps are being primarily sold to catch and monitor Africanized honey-bee swarms in the Southern States. Not that they would not work in your area, but as you mention, there seems to be a reduction in feral colonies and managed colonies due to the current "catch-all" reason, mites. In some parts of the country there are virtually no feral colonies in existence. Most swarms probably will come from strong managed colonies with which the beekeeper has failed to take swarm containment measures.

The swarm traps and lures are relatively inexpensive. It might be fun to set one up and just see.

Question - Do bees really have a sense of taste and if so, why would they collect pollen materials for which they have no real use?

Mabel Marmel
Muscatine, Missouri

Answer - A Dr. von Frisch has done the most experimental work in this field and my answer to your question is based upon his findings. According to Dr. von Frisch, bees are able to distinguish sweet tastes, as we might suspect from their affinity with honey, but can also single out salty, sour, and bitter tastes as well. Honey bees were found to be slightly more sensitive to salty and sour tastes than human, but less sensitive to bitter tastes than humans as they readily took a mixture of quinine and sugar which would be totally disgusting to the human sense of taste. Bees were also found to be sensitive to sugar solutions as low as one or two percent which is considerably more acute than the human sense of taste.

In answer to your question of why bees sometimes gather materials such as sawdust, coal dust, and various animal feeds for which they likely have no real use, Dr. von Frisch also found that hungry or starving bees lowered their thresholds of acceptance in times of need much as a hungry child eats his vegetables better than one with no appetite. Your bees are probably in need of more pollen than local floral sources are supplying. We suggest that you feed a pollen supplement as hunger is driving your bees to collect these materials they will not be able to use.

THIS IS THE GOAL.
GOOD LUCK!